EXTINCTION AND EVOLUTION

WHAT FOSSILS REVEAL ABOUT THE HISTORY OF LIFE

PLATE 1 (See page 47)

***SINOSAUROPTERYX PRIMA*, A FEATHERED DINOSAUR.**

Lower Cretaceous (*ca.* 125 million years). Yixian Formation, Liaoning Province, People's Republic of China.

Feathers were originally thermoregulating devices, helping to maintain body temperatures. But feathers, it turns out, were also precursors for flight: birds are feathered dinosaurs that developed the capacity to fly.

Credit: Staatliches Museum für Naturkunde Karlsruhe.

EXTINCTION AND EVOLUTION

WHAT FOSSILS REVEAL ABOUT THE HISTORY OF LIFE

Introduction by Carl Zimmer

Niles Eldredge

A Peter N. Névraumont Book
FIREFLY BOOKS

Published by Firefly Books Ltd. 2014

First printing

Publisher Cataloging-in-Publication Data (U.S.)
Eldredge, Niles.
Extinction and evolution : what fossils reveal about the real history of life / Niles Eldredge.
[256] pages : col. photos. ; cm.
Includes bibliographical references and index.
Summary: A resource on fossils and the stories they tell of the history of life on earth.
ISBN-13: 978-1-77085-359-1
1. Paleontology. 2. Evolution. 3. Extinction (Biology). I. Title.
560.2 dc 23 QE714.3.E436 2014

Library and Archives Canada Cataloguing in Publication
Eldredge, Niles, author
Extinction and evolution : what fossils reveal about the real history of life / Niles Eldredge.
Includes bibliographical references and index.
ISBN 978-1-77085-359-1 (bound)
1. Fossils. 2. Extinction (Biology). 3. Evolution (Biology).
I. Title.
QE711.3.E43 2014 560 C2014-901154-7

Published in the United States by
Firefly Books (U.S.) Inc.
P.O. Box 1338, Ellicott Station
Buffalo, New York 14205

Published in Canada by
Firefly Books Ltd.
50 Staples Avenue, Unit 1
Richmond Hill, Ontario L4B 0A7

Cover and interior design by Nicholas LiVolsi
Front cover image: National Parks Service (U.S.)

Printed in the United States of America

Produced by Névraumont Publishing Company, Inc.
Brooklyn, New York

CONTENTS

INTRODUCTION

When I was thirteen, my science class got in a school bus and rode across the river into Pennsylvania. The bus pulled over on the side of lonely road. From the shoulder rose a small exposed outcrop, crowned by trees. We were each given a small, strong hammer and instructions to smash the stony slope. Some kids scrambled up the outcrop, where they could hammer in the cool shade. I preferred to keep my feet on asphalt, and so I toiled in the strong sun. I could hear the pinging hammers above me, and sometimes a little chip of rock landed lightly on my head. I was too distracted to care, because my hammer was revealing just the sort of thing we had all come here for: a series of small black ribs emerging from the rock.

As the ribs came into view, I aimed my blows to avoid smashing them, landing my hammer a few inches away on all sides. Eventually, I fractured the rock and the black ribs tumbled out. Leaning my hammer on my left shoulder, I caught it in the palm of my right hand.

The sight of that ribbed rock is still with me, and will probably never leave. Our teacher had explained to us all a few days before what this was, and now I tried to join his words to my vision. This was not a rock—or, at any rate, not just a rock. It was a rock in the shape of a once-living thing.

The thing in my hand was an animal known as a trilobite. The ribs formed a flexible shield across its back. Its eyes were like a pair of speckled jewels. Trilobites became extinct 250 million years ago, but before then they lived in beetle-like abundance. The trilobite in my hand had crawled across sea floors, and then it died and was buried under a slowly thickening layer of mud. Instead of decaying away, becoming a meal for bacteria and scavenging animals, my trilobite had been transformed through a mineralogical alchemy into a stony replica. And it had remained buried in place ever since, through hundreds of millions of years of Earth's history, as dinosaurs emerged on land and then, evolving feathers, took to the air; as the sediments that were its tomb vaulted from the ocean and became dry land, rose higher into serrated mountains that eroded down into gentle hills, which were overrun millions of years later by little hairless bipedal apes swinging hammers.

This was not my first time in the presence of fossils. I had spent plenty of time as a boy standing before the fossils of tyrannosaurs and mastodons in museums. But in those fossil halls, I couldn't be viscerally sure that those fossils had actually been delivered from the Earth. On that hot day standing by that Pennsylvania outcrop, looking at this little cousin of horseshoe crabs I had liberated from the Earth, I knew for sure.

Extinction and Evolution brings that late summer day back to me. Perhaps it's because trilobites have meant so much to Niles Eldredge. He didn't just dig one out of the ground as I did: he's inspected countless trilobites over the course of many years. And in that time, he has been able to do more than just imagine the life of an extinct animal. He could envision the evolution of trilobites over millions of years, through countless generations. The adventures of a single trilobite—its mating, its laying of eggs, its eventual demise as a predator's meal, perhaps, or the victim of a virus—becomes a tiny tile in a grand mosaic. Only by stepping back and looking at the entire mosaic could Eldredge begin to see patterns in the stasis and pulses of trilobite evolution.

Trilobites have much company in the pages of this book. Each image of a fossil has its own particular anatomical beauty, from whorled snail shells to toothy crocodile snouts. But the genius of the theory of evolution is that it joins nature in all its disparity into a single flow. Genes obey the same logic whether they are carried by bacteria or bears. Mutations give rise to variations, and natural selection and other processes can assemble them into new forms of life. New ecosystems can take shape, which then turn back and guide the transformation of the genes in their constituent species. The fact that a species could become aware of billions of years of evolution is astonishing, and we can count ourselves lucky that that species is our own.

Carl Zimmer

PLATE 2

THREE AMMONOIDS DISPLAYED ON A PAGE OF ILLUSTRATIONS OF FOSSIL NAUTILOIDS IN A MONOGRAPH BY JAMES HALL, PUBLISHED IN 1879.

Paleontologists routinely compare new fossil specimens with illustrations of previously studied species.

PLATE 2

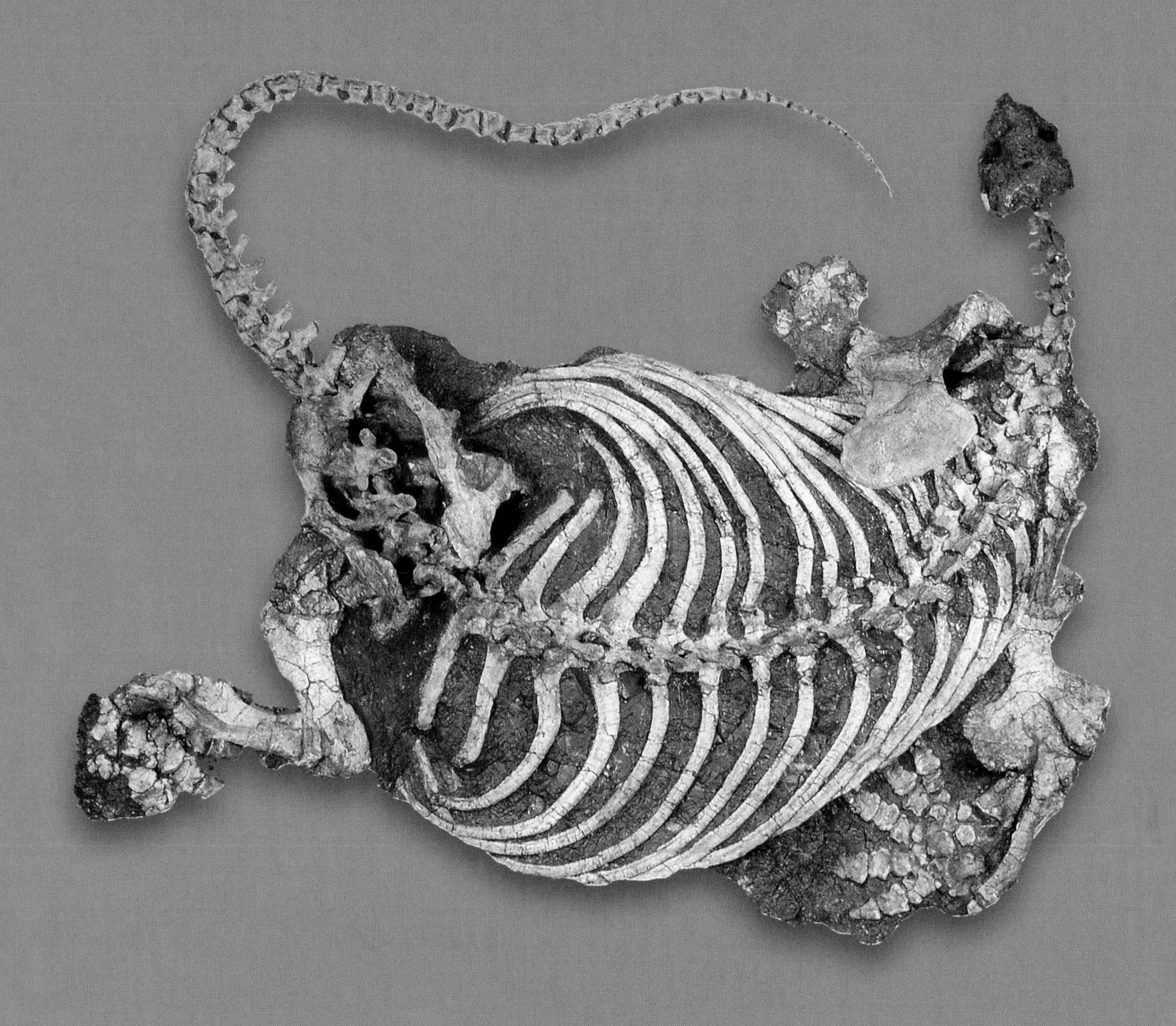

PLATE 3

COTYLORHYNCHUS ROMERI,
A PRIMITIVE SPECIES RELATED TO THE MAMMAL-LIKE REPTILES.

Middle Permian (*ca.* 255 million years). Oklahoma.

Fossils pose questions; we match pictures of fossils with conceptual pictures as we seek to understand the evolutionary process.

CHAPTER 1

THE PAST AS PROLOGUE

PLATE 4

***PSILOPTERA* SP., A JEWEL BEETLE.**

Middle Eocene (*ca.* 47 million years). Messel Shales, Eichstatt, Bavaria.

Colors are rarely preserved in fossils. The Messel Shales are an example of a Lagerstätte (pl. Lagerstätten)—where fossils are preserved in more intricate detail than in more commonly encountered fossiliferous deposits.

Credit: Senckenberg Forschungsinstitut und Naturmuseum.

I have always been fascinated by old things. In school, I preferred ancient to modern history and Latin over French. When I arrived at college, intent upon studying Greek and Latin, I stumbled upon anthropology: what could be more focused on the modern world than the science of human cultural diversity? Yet even in anthropology, my true colors quickly came out: anthropology to me meant far more than the fossil-collecting world of Louis S. B. Leakey than it did the ethnographic studies of Margaret Mead. Sure enough, privileged to play an apprentice role as anthropologist in a Brazilian fishing village in 1963, I became as hooked on digging Ice-Age fossil shells out of the reef, which provided a safe harbor for the village's fishing fleet, as on the details of social organization of the fishing economy itself.

Back at school the following autumn, I took a course in geology—and thereby embarked on a lifetime career trying to make some sense of the fossil record of the history of life. Fossils are inherently geological objects: true, they are remains of ancient life, and my interest in them has always been primarily biological. But first and foremost fossils are of the Earth itself. Accordingly, paleontology, the academic discipline devoted to the study of fossils, has been as much a geological as a biological subject. [Plate 4] Vertebrates—mammals and birds, amphibians, reptiles and the various sorts of fishes—are generally not all that common as fossils (though there are many important exceptions to this rule). Vertebrate fossils have been studied primarily as an outgrowth of comparative anatomy—by zoologists, that is, rather than by geologists.

Invertebrate fossils, in sharp contrast, are common and often extremely abundant constituents of sedimentary rocks. Living on the floors of the shallow seaways, which have usually covered the world's continents over the past billion years, invertebrates are easily trapped and fossilized in sediments and preserved for the ages—and for future paleontologists. Invertebrates are pretty much all the animals you can think of except vertebrates. They include (among many others) such major groups ("phyla") as mollusks (snails, clams, squids and kin), arthropods (crabs and shrimp, insects and relatives), echinoderms (sea urchins, starfish and congeners) and coelenterates (corals and jellyfish), to name but four of the phyla that have made important contributions to (one might say "lasting impressions in") the fossil record. [Plate 5] Geologists quickly discovered that geological time could be subdivided with a fair degree of precision because certain fossils always occur above others, and below still others. The rocks of Earth are "zoned" by their fossilized organic content. Because sediments (muds, silts, sands, plus grains of lime) settle to the bottom of seafloors and lakes, the lowest layers are the oldest. Layers build up, and the animals and plants whose remains are intercalated with

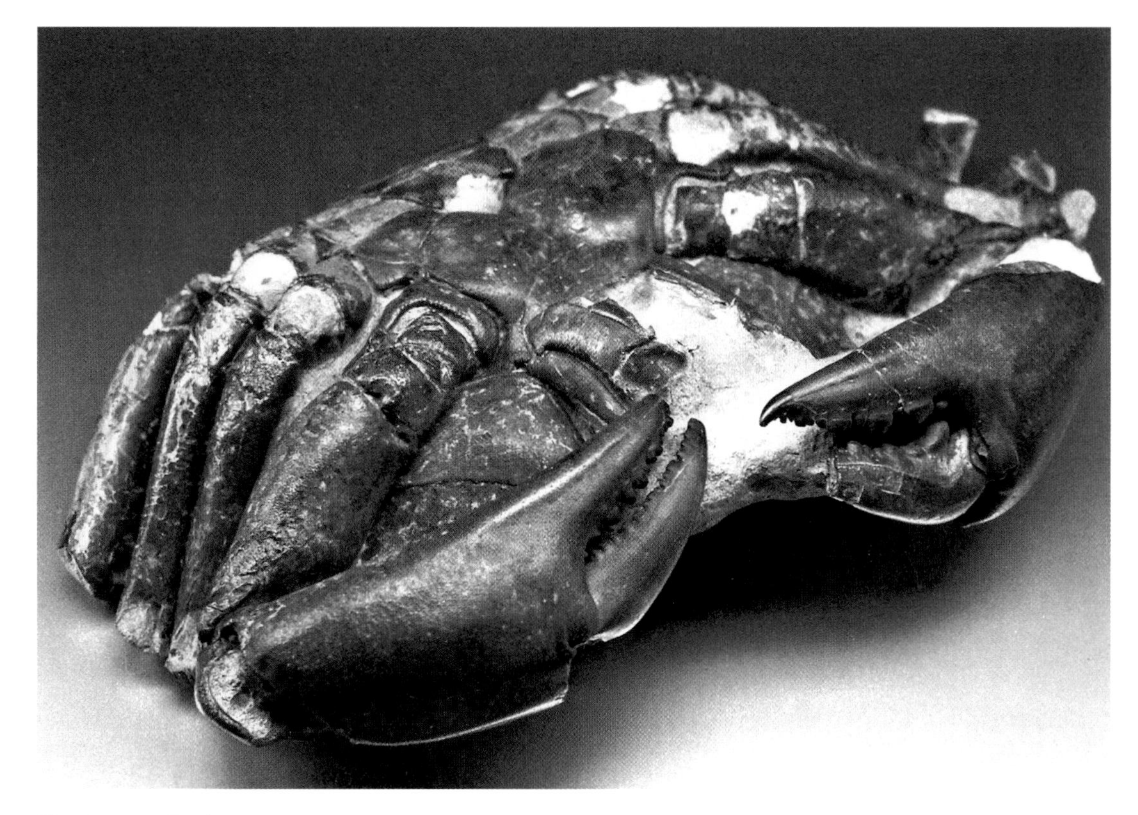

PLATE 5

***GERYON PERUVIANUS*, CRAB VIEWED FROM BELOW.**

Pleistocene (*ca.* 1 million years). Monte Leon, Patagonia.

Crab skeletons are delicate and rarely preserved as exquisitely as this specimen.

PLATE 6

***PLATANUS* SP., LEAF OF A SYCAMORE TREE.**

Eocene (*ca.* 50 million years). Garfield County, Colorado.

By the Eocene, many familiar forms of life had already appeared.

the sediments faithfully record the time they were alive and well on planet Earth.

I have included a chart of geological time for reference here. How geologists use fossils to decipher time is fascinating—but not our central topic in this book. I have been engaged, rather, in trying to piece together how the sequence of fossil forms reveals the course of the evolution of life—and what that history, furthermore, tells us about the very nature of the evolutionary process. Evolution will be the focus of my narrative through the next six chapters—but we must never forget the profound connection between fossils and the *physical* history of Earth.

True to form, of the three major divisions of geological time since the advent of complex life forms, I opted for the earliest—the "Paleozoic." I chose to study trilobites, earliest and one of the most primitive of all known arthropods—and a group now totally extinct. But I have found that I have had to temper my enthusiasm for the purely ancient (and "the older the better" attitude), the better to grasp the meaning of my fossils. I have had to come to grips with the living world—and all that modern biology has so far been able to tell us about the nature of life. Fossils are mere vestiges of formerly living things: their hearts no longer beat, they no longer search for food, nor do they reproduce. We know they all did when alive, but how they did is often difficult to tell—unless we let living creatures come to our aid. [Plate 6]

I sensed my first glimmer of the importance of the living world to interpreting extinct life from my friend and colleague Stephen Jay Gould. We were in graduate school together, he two years ahead of me. Embarking on his research, Steve went in just the opposite direction from my adoration of the Paleozoic. He chose the Pleistocene, the "Ice Age," which began a little more than one and a half million years ago, and only ended 10,000 years ago (some geologists indeed think we are still in the Pleistocene). Steve chose to study the evolution of some land snails on Bermuda: a "microcosm" as his penchant for the *mot juste* led him to call it. The snails were literally landlocked, trapped on Bermuda, and evolved right there—no messy migration in and out to disturb the picture. And, best of all, descendants of those Ice Age fossils are still very much alive in Bermuda. Steve had a living, breathing functioning life system whose origin and history right up to the present was abundantly preserved in the fossilized sand dunes and ancient soils of primordial Bermuda. Extrapolating from what he knew already about the ecological requirements of the living snails, he could understand what happened in the earlier evolution of various fossil snail species as a response to environmental change as the climate fluctuated during the Ice Age.

GEOLOGICAL TIME SCALE

Millions of years before the present	ERAS	PERIODS	EPOCHS	*Duration of Eras (in millions of years)*
.01	CENOZOIC	Quaternary	Recent	66
2			Pleistocene	
5.3		Tertiary	Pliocene	
23			Miocene	
33.9			Oligocene	
56			Eocene	
66			Paleocene	
145	MESOZOIC	Cretaceous		186
201		Jurassic		
252		Triassic		
299	PALEOZOIC	Permian		289
323		Pennsylvanian	Carboniferous	
359		Mississippian		
419		Devonian		
445		Silurian		
485		Ordovician		
541		Cambrian		
4,550	PRE-CAMBRIAN			4,009

= Major Extinction Event

PLATE 7

I have come to appreciate just how intimately our understanding of the past depends upon a sure grasp of the dynamic principles underlying the organization of today's living world. Yet I have by no means abandoned the traditional paleontological perspective, which tells us that, to understand the modern world, we must look to the past to see how things have come to be as we find them today. Therein lies the core of my personal approach to both the history of life and to the study of the evolutionary process. The fossil record has much to tell us about what has happened in evolutionary history. Modern biology has much to tell us about how organisms come to be adapted to their environmental surroundings. It is my goal to conceive and articulate a better match between these two sources of vision than was available when first I set out on my paleontological career. [Plate 7]

Putting the fossil record together with perspectives on the living world is not as easy as might at first glance be assumed. Paleontology had become so divorced from the modern world of biology after the nineteenth century that paleontological theories of evolution often conflicted sharply with the ideas put forward by geneticists. Genetics—the study of the principles of heredity—had made great strides since its inception around 1900. In the exuberant atmosphere that comes with rapid progress in a new scientific discipline, some geneticists actually thought that they could supplant the original Darwinian vision of evolution through natural selection with other, more direct, mechanisms of hereditary change. By the 1930s, though, geneticists had been able to effect a rapprochement between their newly acquired knowledge of genes and Darwin's original explanation of the dynamics of the evolutionary process—a coherent theory called "neo-Darwinism." [Plate 8]

What remained, in the late 1930s and early 1940s, was to achieve a further integration between the neo Darwinian vision and the broad sweep of the history of life. Paleontologist George Gaylord Simpson, who will appear again in our narrative, undertook to put the data of the fossil record together with modern evolutionary theory in the 1940s. He did so boldly and imaginatively—while admitting that the job was "possibly hazardous."

Why hazardous? In his own humorous words, here are Simpson's thoughts, in 1944, on the difficulty of the task of melding genetics and paleontology:

"*Not long ago paleontologists felt that a geneticist was a person who shut himself in a room, pulled down the shades, watched small flies disporting themselves in milk bottles, and thought that he was studying nature. A pursuit so removed from the realities of life, they said, had no significance for the true biologist. On the other hand, the geneticists said that paleontology had no further contributions to make to biology, that its only point had been the completed demonstration of the truth of evolution, and that it was a subject too purely descriptive to merit the name 'science.' The paleontologist, they believed, is like a man who undertakes to study the principles of the internal combustion engine by standing on a street corner and watching the motor cars whiz by.*"

Simpson was writing of the collision of two different perspectives—two different views of the world. He acknowledged (as we all must) that whatever we think about evolution must be consistent with the known principles of genetics. Indeed, he conceded that most of the critical factors of the evolutionary process lie squarely in the domain of genetics. On the other hand, he maintained that whatever we say about evolution must likewise agree with what we know about the history of life as revealed in the fossil record. As we shall see, Simpson had a grasp of an important aspect of the history of life that seemed to him (correctly, in my view) not to square with the standard picture of evolution prevalent in the evolutionary biological theories of his day.

What are these "pictures" of the history of life? At one level, we have the physical evidence itself—gorgeously portrayed in photographs of a rich spectrum of fossil animal and plant life featured in these pages.

PLATE 7

OPHIODERMA EDGERTONI,
A BRITTLESTAR ECHINODERM.

Lower Jurassic (*ca.* 200 million years).
Lime Regis, England.

Paleontologists study living brittle stars to understand fossil species–and fossils species have much to tell us of the history behind living brittlestars.

This is our base line: we want to explain how and why these creatures, and countless other species of life, came into being, why they look the way they do, and why they are no longer with us. These photographs are samples of the primary documents of life's history—and we have no trouble at all seeing them literally as "pictures." [Plate 3, see page 8.]

But words paint pictures too. In the quest to make sense of the fossil record of life's history, to put it together in an ever better fit with the principles of genetics and modern evolutionary theory, we need to see beyond the particulars of this fossil and that: we need to look for general *patterns*. As we shall see, modern evolutionary biology has retained Darwin's original picture of what the fossil record *should* look like if his ideas on the nature of the evolutionary process were correct in all details. The struggle for a better evolutionary theory, from a paleontological perspective, is the story of reconciling Darwin's views (as modified by modern biology, especially genetics) with the actual patterns—pictures revealed by the fossil record—of life's history.

Science is the ongoing struggle to make our ideas conform ever more closely to the patterns which we perceive in nature. Ultimately, our conceptual pictures of nature are part perception and part interpretive explanation. The job is never done: it is each generation's charge to tinker with the picture, trying to bring it into ever sharper focus as more is learned. Collisions are inevitable—as each new generation questions the wisdom of its predecessors, and as different scientists dispute the necessity of altering the picture at all, and if so, precisely what shape the alterations are to assume. But I can think of no more exhilarating a venture than probing ever more deeply into life's past and tinkering with the evolutionary picture. The word pictures in this book supplement the photographic pictures—and document one paleontologist's quest, as it stands so far, to gain a clearer image of evolution.

PLATE 8

***PHAREODUS* SP., A PRIMITIVE TELEOSTEAN BONY FISH.**

Lower Eocene (*ca.* 50 million years). Fossil Lake, Green River Formation, Wyoming.

Without fossils to outline the history of life, evolutionary biology would lack the drama, sweep and pattern of life's history in geological time.

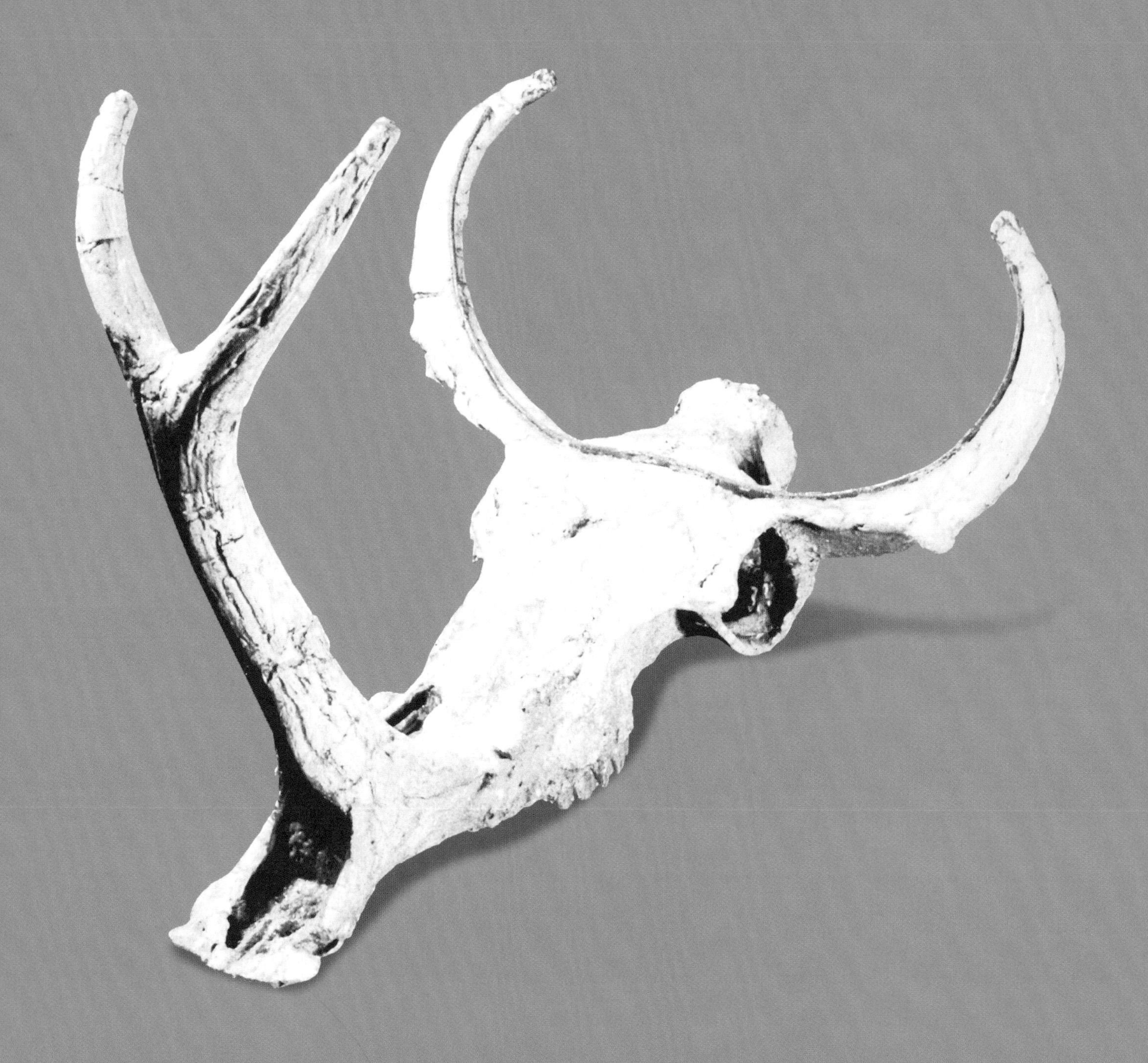

PLATE 9

***SYNTHETOCERAS TRICORNATUS*,**
ARTIODACTYLS (CLOVEN-HOOFED) CAMEL RELATIVE.

Miocene (*ca.* 12 million years).
Clarendon, Texas.

The peculiar horns of this extinct creature probably served for mate recognition—as in the horns of modern-day African antelope.

CHAPTER 2

ADAPTATION

Chaos theory continues to be the subject of much research among mathematicians, meteorologists, physicists, engineers, economists, biologists and philosophers. Even when there seems to be order, chaos is often lurking just below the surface. A weather map of North America, for example, can look fairly simple and orderly–with a low pressure region in the Pacific Northwest (rain in Seattle!), a Bermuda high dominating the center, and another low forming off the Carolina coast. Yet these patterns arise from local aberrations that can neither be measured nor anticipated. Chaos is the nemesis of long-range weather forecasting.

We have just the opposite problem when we confront the awesome vastness of life's history. Here, at first glance all seems to be chaos: we have *trillions* of organisms, *billions* of years and *millions* of species with which to contend. Even if we start with life's humble beginnings as lowly bacteria way back in the dim reaches of early Earth history, we may quickly lose our way as the initial single strand splits into a bewildering array of microbes, fungi, plants and animals.

But we should not despair. There is real order in all this apparent chaos. Life has had a long and complex, but ultimately comprehensible, history. There are patterns repeated over and over again as new species come and go, and as ecosystems form and fall apart (or are ripped asunder). These organizing principles of life's history are the processes of evolution. We apply them to fossils to render that order. And we look at the fossils themselves to tell us directly about that order in the history of life. The lessons are simple, yet reveal profound truths about the very nature of life and its history. The search for understanding is the search for pattern, for order. And the first, and perhaps the most profound, signal of order in organic nature is the notion of adaptation.

DESIGN IN NATURE

No one knows exactly how many species there are alive on Earth right now. There are at least one, and possibly as many as eighty, million different species still on the planet. Add to that the untold millions of species that have already lived and become totally extinct during life's 3.5 billion year history and organic diversity seems almost mind-boggling. Whence this vast array of so many different kinds of creatures?

All organisms are faced with two sets of problems as they live their lives: they must secure some form of energy and basic materials with which they shape their very selves, developing from single cells into complex organisms, growing and keeping their bodies functional. [Plate 10] But organisms also reproduce, an activity commonly calling for specialized structures and behaviors in addition to those necessary to just staying alive. One way to come to grips with that vast array of organic diversity is to realize that there are a lot of different ways to make a living, and even many ways to reproduce. [Plate 9, see page 18] Some animals eat other animals, while others eat plants: carnivorous mammals appear and act quite differently from cud-chewing ruminants. On the other hand, the great panoply of orchid shapes and colors lies purely in the reproductive structures of the flowering plants. [Plate 11]

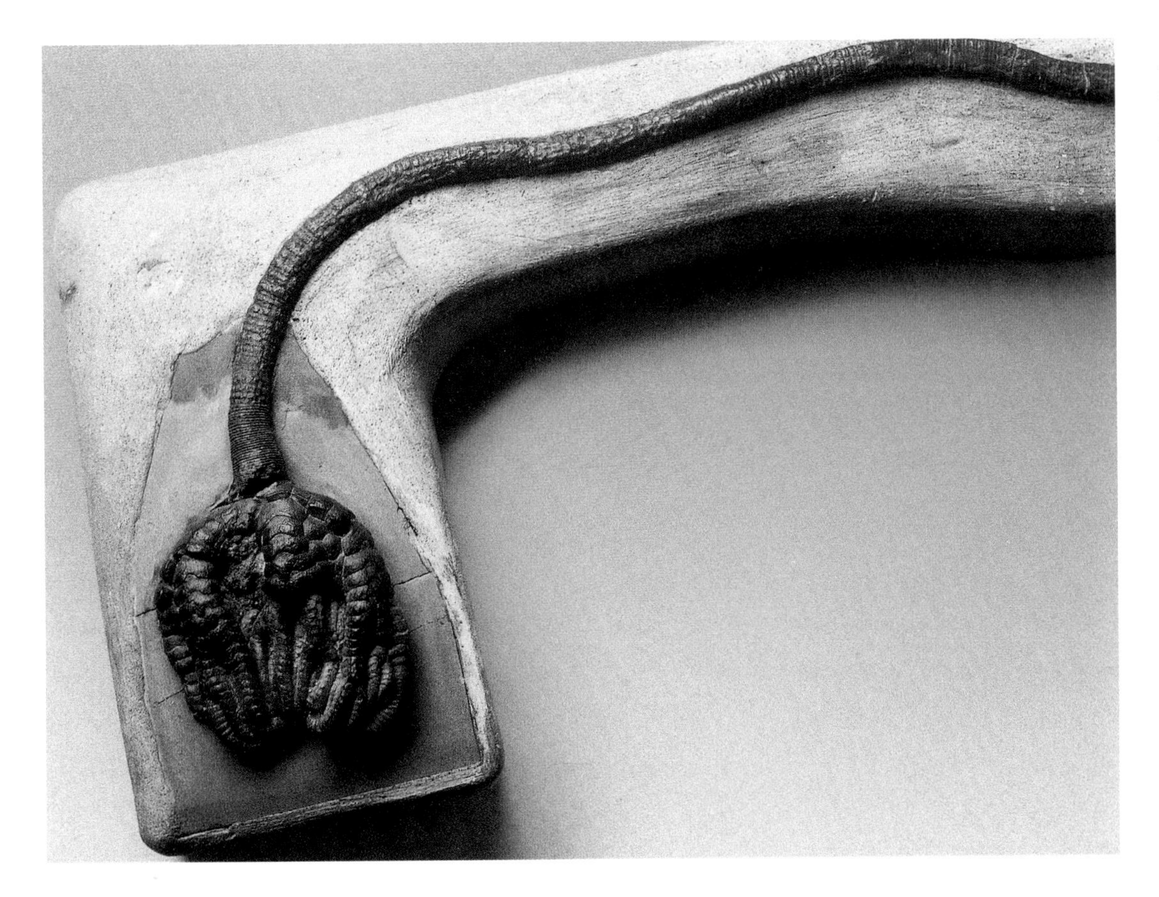

PLATE 10

***FORBESIOCRINUS MEEKI*, A SEA LILY (CRINOID) ECHINODERM.**

Lower Mississippian (*ca.* 340 million years). Crawfordsville, Indiana.

Beautifully adapted to life on the seafloor, crinoids use their feathery "arms" to filter microscopic food particles from seawater.

PLATE 11

ROOT STRUCTURE OF *EUCALYPTOCRINITES SP.*, A CRINOID ECHINODERM, OR SEA LILY.

Middle Silurian (*ca.* 425 million years). Waldron, Indiana.

Rooted to the seafloor, crinoids are superficially plant-like, but are really animals related to starfish and sea urchins.

People have long been aware that organisms seem extraordinarily well-suited to performing their "roles" in nature. The roles all involve capturing energy and material to stay alive. Almost as a side venture, organisms reproduce using available energy and additional structures and behaviors "dedicated" to the purpose. Especially in cases where the energy-capturing devices seem particularly narrowly channeled, it seems rather obvious that organisms are well equipped to perform their economic and reproductive tasks. A house sparrow hopping around on the ground picking up seeds might not seem strikingly well-tuned to its habitat (though their stout and rather stubby bills are indeed ideal for cracking those seeds open). But a walking stick insect, camouflaged to look for all the world like a twig, and able to stay motionless, frozen on its branch, seems a marvel of economic engineering. Such species are highly specialized, while house sparrows are generalists able to live in a broad range of habitats. This distinction between specialists and generalists will be crucial as we seek to explain why some groups of organisms evolve—and become extinct—more quickly than others.

We have come, in short, to the perception that there is design apparent in the living world. Organisms are suited to their environments. Creatures living below the waves extract oxygen directly from water (no mean task), while those of us on land extract the same substance from air. But some land-dwellers do both—as gilled larval amphibians are wont to do. And many adult amphibians continue to extract 0_2 directly from the moisture clinging to their skins. [Plates 12, 13] We see that there is some order to the vast array of different kinds of living things.

PLATE 12

FROG (CLASS AMPHIBIA), YET TO BE GIVEN A SCIENTIFIC NAME AND TAXONOMIC CLASSIFICATION.

Lower Eocene (*ca.* 52 million years). Fossil Lake, Green River Formation, Wyoming.

The hind legs of frogs are, of course, highly modified for prodigious jumping—an evolutionary adaptation. In contrast, the fore- and hind-leg of salamanders are much more nearly the same size, remaining more or less in the same proportions as ancestral amphibians.

Credit: Lance Grande, The Field Museum.

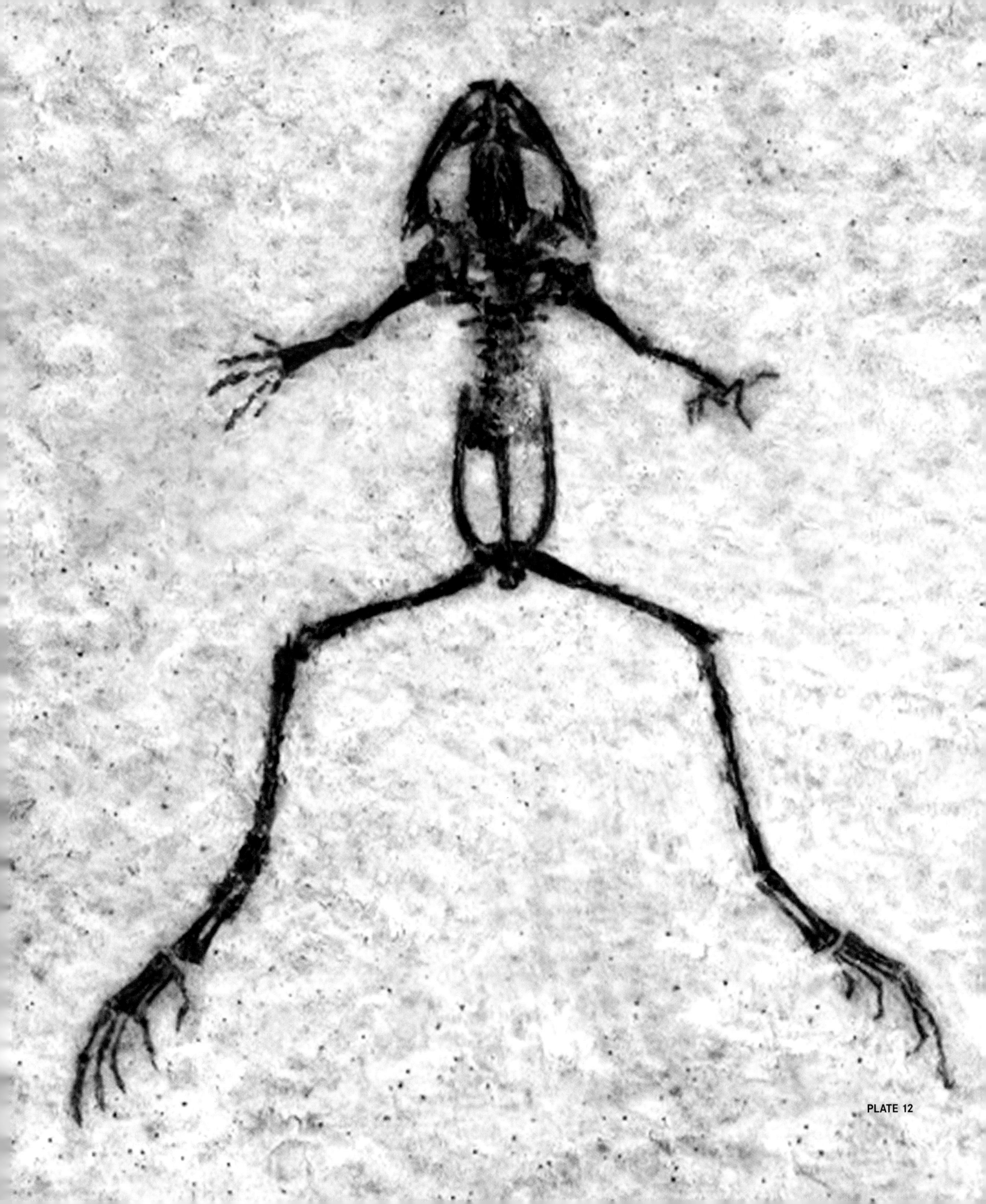

PLATE 12

PLATE 13

***CHUNERPETON TIANYIENSIS*, A SALAMANDER.**

Middle Jurassic (*ca.* 161 million years). Daohugou Beds. Ningcheng County, Nei Mongol, People's Republic of China.

This extinct salamander closely resembles living salamanders, including the endangered Asian giant salamander.

Credit: Ke-Qin Gao, Peking University.

Whence this order and apparent design in living beings? How is it that organisms generally seem so well suited to the exigencies of making a living in the very habitats in which we find them? [Plates 14, 15, 16] As western civilization awoke from the conceptual slumber of the middle ages, the curious began considering how the natural world is put together. Karl von Linné, a seventeenth-century Swedish naturalist better known to us as Linnaeus, perceived a pattern of similarity linking up absolutely all living things: dogs are more similar to cats, say, than they are to deer. But all three have hair and mammary glands and so belong to the same larger group—the mammals. And mammals, birds, reptiles, amphibians and fishes all have backbones and other structures typical of "vertebrates." [Plate 17] And so on: there is a natural order of hierarchically nested groups of living beings. Not only are organisms well suited to their habitats (as if by design), but there is a discernible and systematic pattern of similarity linking all organisms together.

PLATE 14

***CANIS DIRUS*, CLOSE-UP VIEW OF THE DIRE WOLF SPECIMEN OF PLATE 15.**

Dissimilar as they may seem, dogs and cats are closely related, and their skeletons reveal many common features.

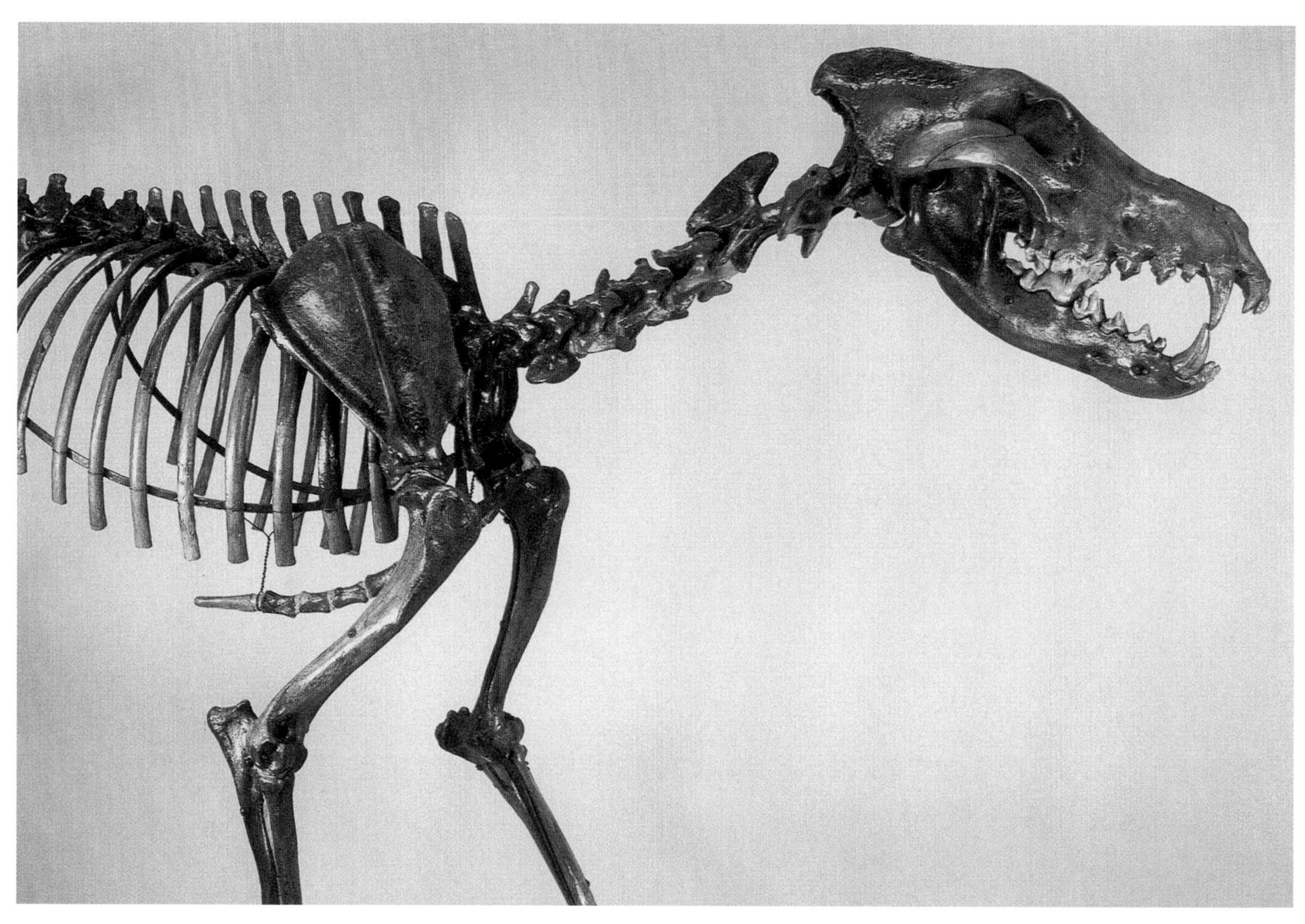

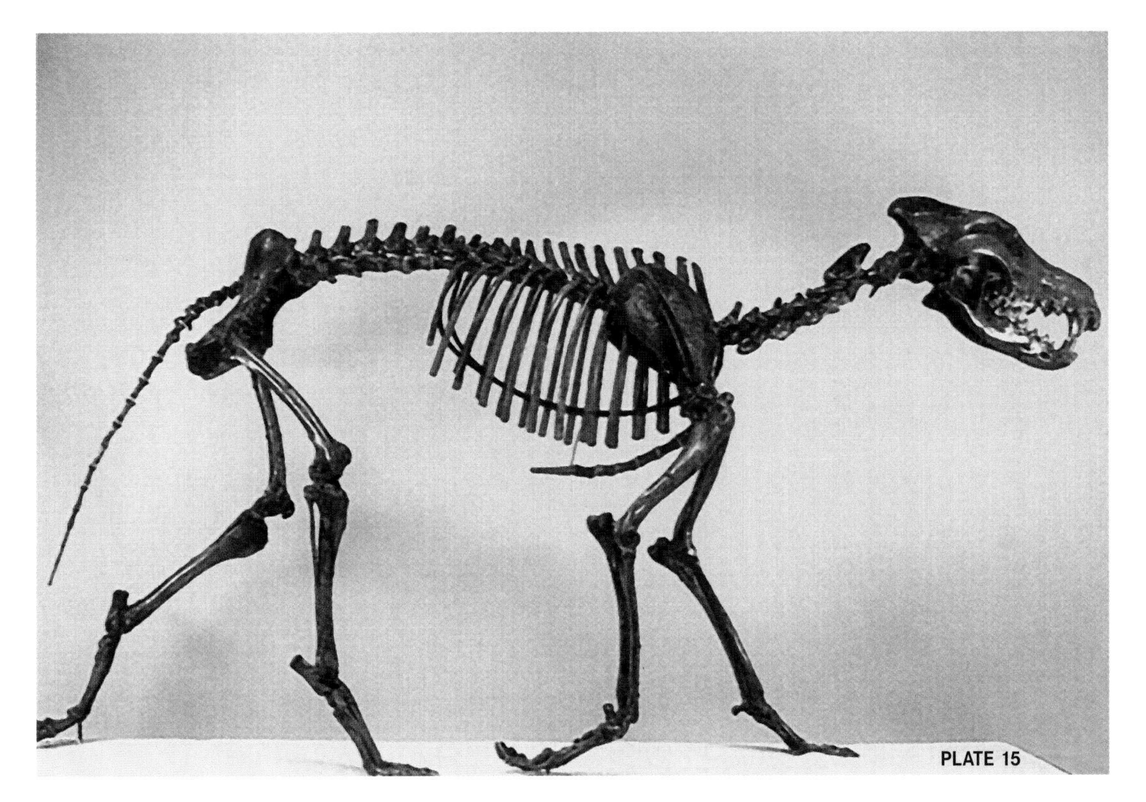

PLATE 15

***CANIS DIRUS*, THE DIRE WOLF.**

Pleistocene (*ca.* 25,000 years). Rancho La Brea, Los Angeles, California.

The dire wolf was well equipped for life as a carnivorous mammal.

PLATE 16

***HOPLOPHONEUS PRIMAEVUS*, AN EARLY FORM OF SABER-TOOTHED MAMMAL CLOSELY RELATED TO TRUE CATS.**

Oligocene (*ca.* 30 million years). South Dakota.

All carnivorous mammals share a common pattern of shearing teeth in their jaws.

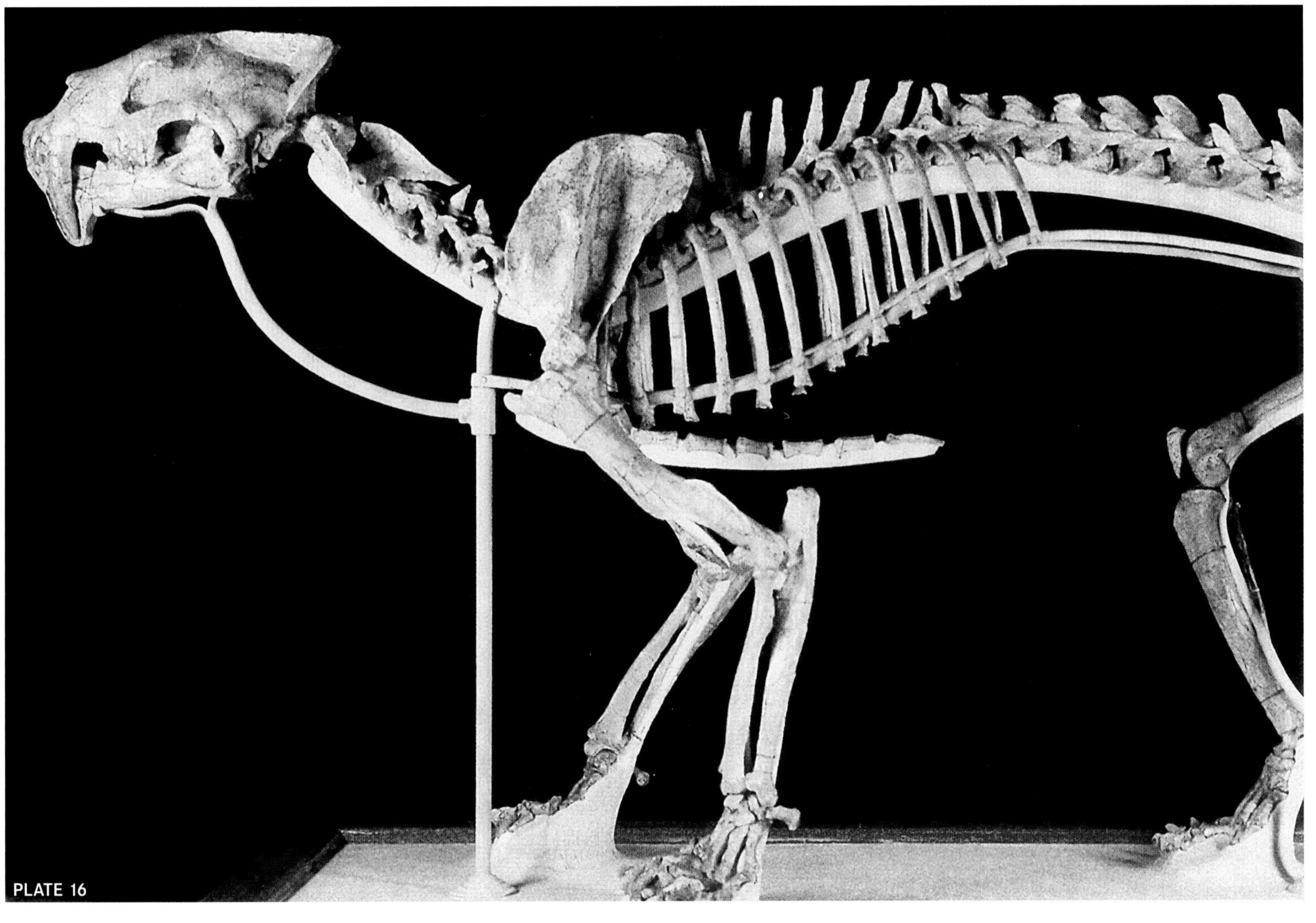

PLATE 17

***ERYOPS MEGACEPHALUS*, A GIANT PRIMITIVE AMPHIBIAN.**

Lower Permian (*ca.* 270 million years). Wichita Basin, Texas.

Amphibians are land-dwelling vertebrates that lack the amniote egg, with its inner membranes, shared by reptiles, birds and mammals.

The early naturalists confronting patterns of biological forms (including the fossil record—Linnaeus named many species of fossils along with living organisms) were faced with a ready-made answer to the questions: Whence this diversity? Whence this natural design? Whence this ordered pattern linking all living things? The answer: God the Creator. Many of the savants of the Renaissance were themselves clergymen. Those who were not were nonetheless typically pious. As natural science blossomed forth in the waning days of the eighteenth, and throughout the first half of the nineteenth, centuries, rapid strides were made in geology, biology and that hybrid between the latter two: paleontology—the study of the fossilized remains of ancient life. But, though chemistry and physics could proceed without contending with Biblical revelation, natural history could not: the savants of ancient times who had written the various documents that have come down as our Judaeo-Christian Bible had considered all these questions before, and their answers are there for all to see in the opening verses of *Genesis*.

William Paley, canon of St. Paul's cathedral, was perhaps the last great spokesman for the biblical interpretation of the facts of natural history. In his *Natural Theology* (1802), Paley posed his famous analogy between the mechanism of a watch and the design we all agree is evident in living nature. The intricate internal workings of a pocket watch could never have been spontaneously self-assembled. Rather (or so Paley argued), only intelligent design and diligent fabrication could possibly make a watch. A watch implies a watchmaker. So, too, (Paley went on) does the design apparent in each living creature, with its intricately interconnected and interdependent parts, imply a Creator. As the early twentieth century poet Joyce Kilmer said, "only God can make a tree."

This, the "creationist" position, was the accepted explanation for why there are so many different kinds of organisms (because there are so many different environments in which to live), why organisms appear so well suited to their environments (that's the way the Creator fashioned them) and why there is a similarity that interlinks absolutely all of life (creationists often argue that God in effect used a blueprint, from which different body plans were derived). Latter-day creationists still use Paley's watch and other pre-Darwinian creationist arguments. But the speculative naturalists who first produced these elaborate schemes to reconcile the facts of nature with the old Biblical stories rapidly gave way, in the mid-nineteenth century, to an entirely new breed of successors who sought natural, rather than supernatural, answers to these three questions. At the head of this new breed, of course, stood Charles Robert Darwin, the man who, in a single (if protracted) stroke, finally weaned western thought from its ages-old dependence on supernatural explanations for natural phenomena.

CHARLES DARWIN, DESCENT WITH MODIFICATION AND NATURAL SELECTION

Darwin's epochal *On the Origin of Species by Means of Natural Selection* (1859) begins most unpromisingly: he tells his readers the details of his experiences as an apprentice pigeon fancier. No sweeping details of life's history, no pageant of the fossil record, no recital of the facts of life's stunning diversity in the wild. Just pigeons.

It was Darwin's task to convince his readers that the grand questions of biology and paleontology could be answered by simple recourse to processes we see going around us everyday. When, as a young man in 1831, Darwin set out for a five-year voyage on the *H.M.S. Beagle*, the only book he took with him was the just published first volume of Charles Lyell's *The Principles of Geology*. It had been Lyell's greatest accomplishment to persuade his peers that a rational approach to a study of Earth and its history began with the assumption that modern day processes were operating as well in the past. We can use our knowledge of the present to understand the past—a dictum originating with the Scottish physician-farmer-naturalist James Hutton in his *Theory of the Earth*, and developed further when Lyell came along.

Darwin, of course, had serious business in mind when he wrote of his pigeon breeding experiences. He was proposing a solution to the three key questions on diversity, design and interconnectedness of all living things. For he saw the interconnectedness—that haunting pattern of similarity that links up all living creatures in a progressive, hierarchically nested manner—as a primal clue. Just as Earth, it was coming to be realized, had had a terribly long history (far in excess of the 6,000 or so years admitted through Biblical computations), so might life have had a great history as well. Others before Darwin (of his generation, and, intermittently, stretching back to the roots of western thought) had come to much the same conclusion. They realized that a process of "descent with modification" (as Darwin called it) would automatically produce just that nested pattern of similarity that Linnaeus had focused on. If, as time passes, a lineage is split into many different lines of descent, then, as history goes on, if new features arise in one lineage, they will be inherited only by later descendant members of that particular lineage. The other lineages will not share that new feature. Because new features are added as time goes by, the more recently two lineages share a common ancestor, the more features they will have in common. That would give a good, natural reason why cats and dogs are more similar to one another than they are to deer—and why dogs, cats and deer are more closely related to (or recently diverged from) one another than any of them are to, say, birds. [Plate 18]

PLATE 18

***EURYPTERUS LACUSTRIS*, A EURYPTERID, OR WATER SCORPION, RELATED TO HORSESHOE CRABS; VENTRAL (BOTTOM) VIEW OF A FEMALE SPECIMEN.**

Upper Silurian (*ca*. 415 million years). Buffalo, New York.

Anatomical structures on this fossil are similar to those seen in horseshoe crabs and spiders, evidence that all have descended from a single common ancestor.

PLATE 18

PLATE 19

RHYNCHOTHERIUM SHEPARDI,
A RELATIVE OF TRUE MASTODONS.

Upper Pliocene (*ca.* 3 million years).
Mt. Eden, California.

Though common in the fossil record, no elephant species has ever multiplied and taken over Earth!

But how would that process work? It was Darwin's answer–*natural selection*–that simultaneously persuaded the thinking world that life has indeed had a long evolutionary history, and provided a naturalistic counterpart to Paley's Creator as the source of all that apparent design in nature. Which takes us back to Darwin's pigeons.

Breeders for at least the last 10,000 years have realized that qualities of organisms can be enhanced in ways humans think desirable: lambs can be produced that grow thicker coats of wool faster than their ancestral stock; milk yields in cattle can be increased; and crop plants can be so highly modified that Brussels sprouts, kale, cabbage, broccoli and cauliflower are all just different lineages derived from the same ancestral "wild cabbage" stock, *Brassica oleracea*. The process is *artificial selection*, and works simply by letting only those few members of a single generation in which the desired trait is best expressed do the breeding for the next generation. All one needs is a keen eye, diligence and patience.

Darwin took from his breeding experiences a heightened awareness that organisms show variation. No two pups in a litter are exactly the same. And he also realized that selection would work only if traits were heritable: higher milk yields in some cows do show up in their descendants. It is a signal fact that Darwin's understanding of the mechanisms of heredity was incorrect. But his concept of selection depended only on the occurrence of heritable variation, whatever its underlying cause would prove to be.

Artificial selection, though, is but a variant version of Paley's watch: human selective breeding can force cattle to produce more milk (or, indeed, produce domestic cattle from wild bovines in the first place). Is there a process in nature similar to this patently human activity? Darwin saw that there is–an insight shared by several others, including naturalist Alfred Russel Wallace (in a malarial dream while in what is now Indonesia). Darwin and Wallace were aware that variation is also present in natural populations in the wild. Though they had no possible way of testing the proposition, they also assumed (logically enough) that variation is as heritable in the wild as it is under domestication.

But what in nature serves the role played by the breeder's hand in artificial selection? The answer came from a clergyman–the Rev. Thomas Malthus, whose 1798 *An Essay on the Principle of Population* argued that resource limits in turn limit human population size. Where population growth outstrips food resources, famine, disease and death soon follow. Darwin took a very general message from Malthus. Without a check on population growth, the natural increase of even such slow breeders as elephants would be exponential. Darwin, starting with a single pair of elephants, and allowing a rate of one offspring every 10 years for 60 years, calculated that there would be nearly nineteen million elephants descended from that first pair in the short span of 750 years. But the world is clearly not wall-to-wall elephants: there simply *must* be some factors limiting elephant population growth. Darwin pointed to predation, disease, other environmental hazards and the most important of all–limitation of suitable food. [Plate 19]

Eureka! Malthus had given him that final connection: not all organisms produced each generation in the wild can possibly survive and reproduce. As a rule, those best suited to life's exigencies, to the dual game of making a living and reproducing, will out-survive and out-reproduce their less well-endowed fellows. And their offspring will tend to inherit those very features that conveyed success on their parents. While the stock will tend to remain the same as environmental conditions remain stable, Darwin asked his readers to imagine what would happen when conditions change—as was well known even in the mid-nineteenth century inevitably happens through geological time. Climates become warmer, or colder; seas rush in where forests once stood (accounting for the great coal deposits that fueled the fires of the Industrial Revolution, then still very much in progress).

Darwin's answer: as conditions change, the balance will be tipped in favor of those variants who were less successful under the old regime, but who do a whole lot better under the new environmental setting. Now, under changed conditions, it is a new ball game, and selection will start favoring alternative variants, changing, as each generation goes by, the overall complexion of an entire species. Voila! A *natural* selective process, analogous to the human pursuit of artificial selection.

With natural selection, Darwin had his explanation of design in nature: the traits of organisms, over generations, become modified either to match changed environmental conditions, or to match environments they have newly entered. [Plate 20] This is the process of adaptation: modification of structure to meet environmental conditions. Natural selection would shape the modifications that, in retrospect, over evolutionary history, produce that array of similarity that interlinks all living things. Of the creationists' three answers, he kept but one: there are so many diverse kinds of living things indeed because there are so many different ways of making a living in the natural world. But organisms have assumed those shapes, playing those varied roles (or, in modern parlance, filling those varied niches) because they have been shaped, not by a Creator, but by natural selection.

PLATE 20

***RAFINESQUINA*, SP. BRACHIOPOD ENCRUSTED BY *MONTICULIPORA* SP., A BRYOZOAN, OR MOSS ANIMAL.**

Upper Ordovician (*ca.* 440 million years). Olive Branch, Ohio.

Natural selection can mold the features of organisms belonging to two unrelated species, enabling them to live in close association: the phenomenon of commensalism.

Darwin painted a picture of the history of life as a slow, steady, rather gradual and even progressive dance between organism and environment. Evolution is the response of organisms to environmental change. Individual organisms do not change; rather, with each generation, natural selection increases the frequency of traits most favorable to survival in the conditions of the moment. Evolution is first and foremost the study of the physical features of organisms: their use and bearing on organismic survival; and their modification for new uses, or even their loss, as conditions change. Organic design and diversity both derive from a natural tendency of life to survive, to make the most of the conditions and resources available at the moment.

Artificial selection can effect some impressive results over a few short generations. But it is always limited: by the available variation within each generation, and by the lifetimes of the breeders—ultimately by the length of human selective breeding programs, still barely 10,000 years old (though genetic engineering offers to speed things up considerably). Think, Darwin mused, of what can happen over billions of years! Nothing less than the entire richly diverse panoply of life—nearly one hundred major groups (phyla) over the past 3–5 billion years.

Darwin's selective mechanism was a genuine discovery. Natural selection is an ineluctable law of nature. It is the effect that relative economic success has on reproductive success. Using selection to model the patterns of the history of life, to tell us why organisms have come to be so diverse, works beautifully to what might be called a "first order of approximation." We will see in ensuing chapters how the actual history of life is considerably more complex than Darwin's simple picture of gradual transformation matching progressive environmental change through time. But adaptation lies at the very heart of the evolutionary process. It is the central theme whenever we look at any organism—certainly including the great array of extinct forms of the fossil record.

CONVERGENT EVOLUTION: ADAPTATION IN ACTION

The fossil record is a laboratory of past evolutionary "experiments" whose results lie frozen in the rocks. Paleontologists approach their fossils with the aim of deciphering their evolutionary history. We have the further goal of additional further light on the very nature of the evolutionary process itself.

The most graphic testimony to the importance of adaptation in evolutionary history is the phenomenon of convergence: organisms coming from separate evolutionary lineages may display remarkable similarities despite having evolved separately from ancestors that were quite unalike and that lived sometimes millions of years apart and in diverse geographic locations. The classic example is the wings of birds, bats and those Mesozoic Era flying reptiles, the pterosaurs. Birds, bats and pterosaurs evolved from separate ancestral, four-legged, ground dwelling stocks: in the case of birds, from dinosaurs; pterosaurs from other reptilian forebears; and bats from a primitive mammalian group that includes dogs, whales, cows, and some insectivores. [Plates 21, 22, 23]

PLATE 21

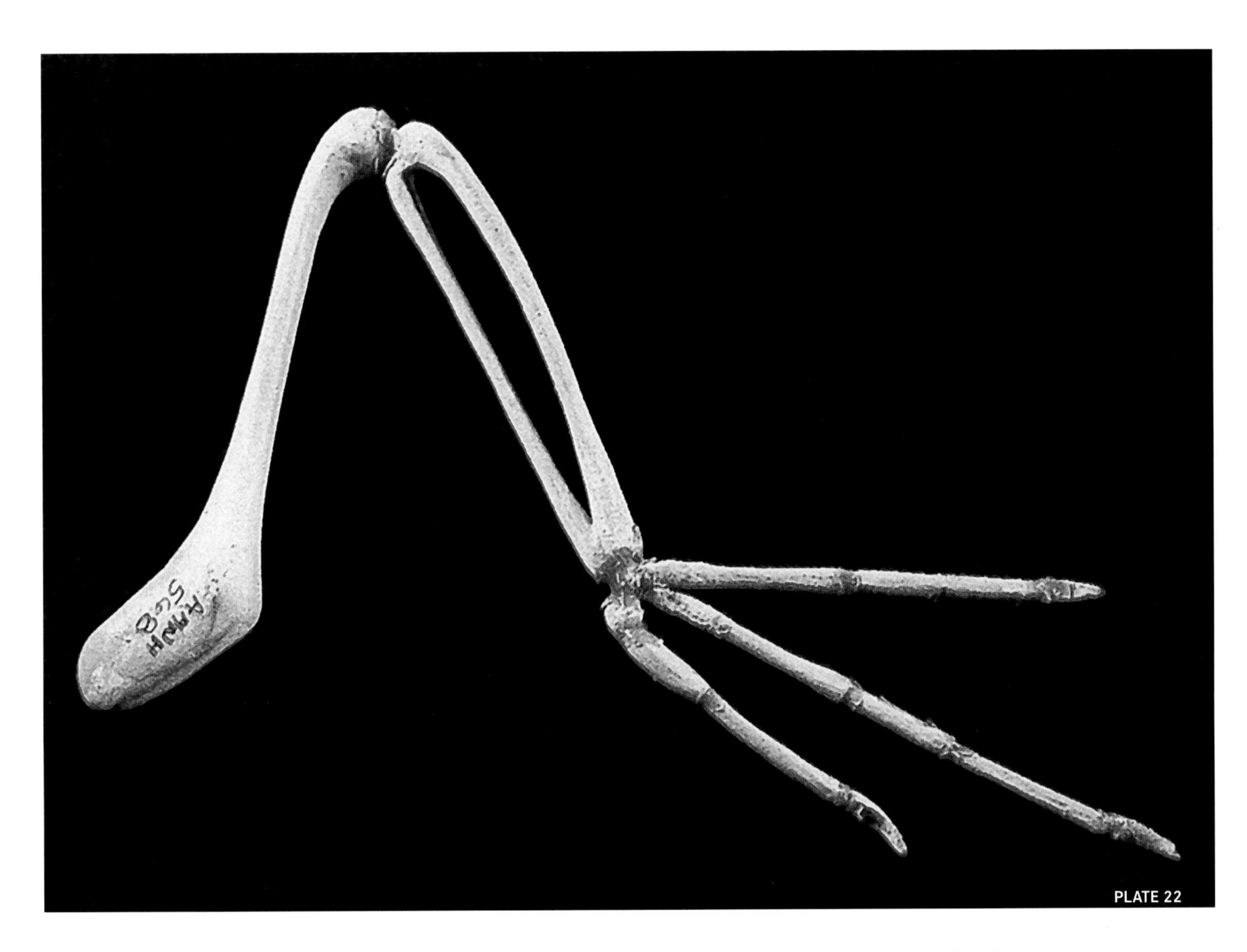

PLATE 21

***EOCYPSELUS ROWEI,* A RELATIVE OF SWIFTS AND HUMMINGBIRDS.**

Lower Eocene (*ca.* 52 million years). Fossil Lake, Green River Formation, Wyoming.

Paleontologists have concluded that this exquisitely preserved bird fossil probably comes from the stem group that gave rise to both hummingbirds and swifts. It could neither hover like a hummingbird, nor dart and swoop like a swift—both of which were later adaptive innovations from something very much like this fossil species.

Credit: Lance Grande, The Field Museum.

PLATE 22

SKELETON OF THE WING OF A MODERN BIRD.

Though similar in overall design and effect, the skeletal structures of the wings of bats (see Plates 23 and 127), birds and pterosaurs are all quite different.

PLATE 23

***RHAMPHORHYNCHUS MUENSTERI*, A PTEROSAUR, OR FLYING REPTILE.**

Upper Jurassic (*ca.* 150 million years). Solnhofen Limestone, Bavaria, Germany.

The skeletal support of the wing of this pterosaur is beautifully preserved in this specimen.

PLATE 24

WORTHENIA TABULATA, **A GASTROPOD MOLLUSK, OR SNAIL.**

Upper Pennsylvanian (*ca*. 295 million years). Jacksboro, Texas.

The species is common throughout the eastern, central and southwestern United States—wherever sediments preserve traces of the extensive Pennsylvanian seas that flooded the continental interior.

PLATE 25

ANANIAS WELLERI,
A GASTROPOD MOLLUSK, OR SNAIL.

Upper Pennsylvanian (*ca*. 295 million years). Gunsight, Texas.

This species differs from *Worthenia tabulata* (Plate 24) only in the structure of the narrow band on the outer edge of each whorl; together, the two species offer an example of convergent evolution.

My very first scientific paper dealt with this phenomenon of convergence. As a student, I was assigned the problem of unraveling the early evolutionary histories of two species of Pennsylvanian (Upper Carboniferous) Period snail species who resembled one another so closely that hardly anyone bothered, or was even able, to tell them apart. Yet the telltale vestiges of their separate evolutionary histories are plain enough to the initiated eye. Both species belong to a large group of primitive snails—the "pleurotomarians"—that were especially diverse and abundant in the Paleozoic Era. A few of these pleurotomarians survive today, in the form of the large "slit-shells" of deeper marine waters, and the abalones, which dwell in the shallows of the rocky intertidal zone. The slit in the shells of these snails provides a channel for the exhalation of wastes and water that has already passed over the gills. As the snails grow, the slit is gradually filled in by shell material. In my species *Worthenia tabulata* [Plate 24] and *Ananias welleri* [Plate 25], the clue to their separate histories lay simply in the different designs on the surface of their very narrow slit-fillings. Otherwise, the two species, often found lying side by side in the rocks, were identical.

Both species lived for at least 10 million years. And it was no accident that they resembled each other so closely: for as one would vary a bit, becoming slightly taller and narrower through time, so would the other. Later, both species would revert to the lower, wider shape each had had earlier. One species seemed to be actually keeping up with evolutionary changes in the other—though which one was tracking which is impossible to tell.

Convergence tells us something of the limitations on design that govern the evolution of shape and size through time. There are often but a few, perhaps even a single, best ways to design an organism to perform a certain function. Some limitations are purely mechanical—or even simply mathematical. For example, placing identical objects close together over a surface is best accomplished if each object, or cell, is six-sided, which yields optimal packing. Thus many colonial invertebrates, such as coral and sponges, have six-sided walls. The compound eyes of crustaceans and insects are likewise often six-sided—as are the cells of honeycombs, which, though not themselves organisms, are structures constructed by organisms—honey bees.

The most spectacular cases of convergence involve organisms of two or more different phyla, and embrace literally hundreds of millions of years. My favorite illustration of convergent evolution continually shaping entirely different organisms into similar shapes to play analogous roles virtually throughout the entire span of invertebrate life comes from reef environments. Sponges and corals, major representatives of two very different phyla, have been building reefs for 500 million years now. [Plate 26] And though the cast of characters constructing, or living in the interstices, of reefs has changed greatly over time, typical reef niches have persisted throughout—with different organisms playing analogous roles at different periods of Earth history.

Modern corals are often colonial in form, but today, and going way back into geological times, there are many examples of coral species whose organisms live solitary lives. Single corals are typically conical shells ("calyces") of calcium carbonate. The horn corals of the Paleozoic are perhaps the best known of all. Mesozoic and Cenozoic Era conical corals are not descendants of the Paleozoic species: rather, they evolved independently, *convergently,* after the Paleozoic horn corals had become extinct. [Plate 27]

That two groups of corals independently adopted the simple conical shape for solitary organisms is interesting, but hardly astonishing. After all, the two groups are both corals, simple-tissued creatures with tentacles straining nutrients directly from sea water. One might imagine the same shapes evolving time after time in distant branches of the same large group of organisms. But what transcends the conical-shape story into a major epic of convergent evolution is that there are many other kinds of invertebrate organisms that, like corals, sit on (or are rooted to) the sea floor, filtering nutrients out of sea water. Sponges, for example, also tend to adopt the conical shape. But far more spectacularly, two groups of complex, advanced multicellular animals—brachiopods and mollusks—have also adopted a superficial yet striking coral-like appearance.

Brachiopods are the common "shellfish" of the Paleozoic. They look a bit like clams (both brachiopods and clams have two shells, or "valves"), but unlike clams, brachiopods are not mollusks. They sit there on the sea floor using a long feathery tentacle system to do their filtering. Clams, instead, use gills. Yet their lifestyles are rather similar.

In the extensive reefs formed in the Permian Period, the last division of the Paleozoic, brachiopods were abundant. Some of them (the richthofenids) became adapted to the reef environment by adopting a distinctly coral-like mode of growth: the lower valve became deeply conical, while the upper was reduced to a small flap, a sort of protective cap. [Plate 28] Some 150 million years later, the same phenomenon occurred again—but this time it was the clams

PLATE 26

***HEXAGONARIA PERCARINATA,* A COLONIAL RUGOSE, OR HORN, CORAL.**

Middle Devonian (*ca*. 380 million years). New York State.

Both modern and extinct corals often live colonially, with many individuals attached to one another.

that converged on corals. [Plate 29] Clams had come to dominate bivalved-life in the Mesozoic, following the severe cut-back of brachiopods in the mass extinction that ended the Paleozoic. So, in the Cretaceous Period, again in reef settings, we find rudistid clams. Some of them are several feet long—truly immense by clam standards. Once again, one of the valves has become deeply elongated and drawn out into a conical shape. The upper valve is simply a protective cap—an example of part of the molluscan world secondarily adopting a form and mode of life (*i.e.*, in reefs) that had been exploited by corals and sponges since the dawn of complex life nearly 600 million years ago.

Convergences such as these are windows into the adaptive process: because there are a limited number of ways of making a living as a shelled filter-feeding organism living in a reef environment, we see similar shapes evolving time and again in different groups. [Plate 31] But the same underlying principle of adaptation through natural selection that can make unrelated groups look so similar is also the basis for all the tremendous past and present diversity of form in the living world. [Plate 30] The very core of evolutionary thinking ever since Darwin has been the idea of adaptation: that organisms in general are closely fitted to the demands of their environments. Natural selection modifies the features of organisms over the generations, the better to hone their fit to the environments, and the better to survive as environments, inevitably, change through time. Adaptation and natural selection are as central to modern evolutionary thinking as they were in Darwin's time. There may be limited ways to make a certain kind of living—but there are also many different kinds of livings to be made, many different ways to slice up the environmental pie. Adaptation through natural selection tells us in general terms how and why that diversity has evolved. It is our first and most essential milepost along the way towards understanding life's history.

PLATE 27

***ZAPHRENTIS PROLIFICA*, A RUGOSE CORAL.**

Middle Devonian (*ca.* 385 million years). Falls of the Ohio River, Kentucky.

Some species of rugose (horn) corals lived singly, adopting the same shape seen in many other, completely unrelated groups.

PLATE 28

***PRORICHTHOFENIA PERMIANA*, AN UNUSUAL BRACHIOPOD.**

Upper Permian (*ca.* 255 million years). Glass Mountains, Texas.

This reef-dwelling brachiopod at first glance looks very much like a rugose coral.

PLATE 29

***HIPPURITES RADIOSUS*, A CLAM.**

Upper Cretaceous (*ca.* 70 million years). Dordogne, France.

This clam, also an inhabitant of reef environments, also looks superficially like a coral.

PLATE 30 **(See page 46-47.)**

***PSITTACOSAURUS GOBIENSIS*, SKULL OF PARROT-BEAKED DINOSAUR NEXT TO THAT OF LIVING MACAW.**

Lower Cretaceous (*ca.* 110 million years). Suhongtu, Gobi Desert, Inner Mongolia, People's Republic of China.

Evolutionary adaptation is a form of natural engineering of body parts to perform particular function. Parrot-like jaws have evolved in birds and reptiles, but also in fishes and other groups. (See Plate 155.)

Credit: Paul Serano, University of Chicago.

PLATE 27

PLATE 28

PLATE 29

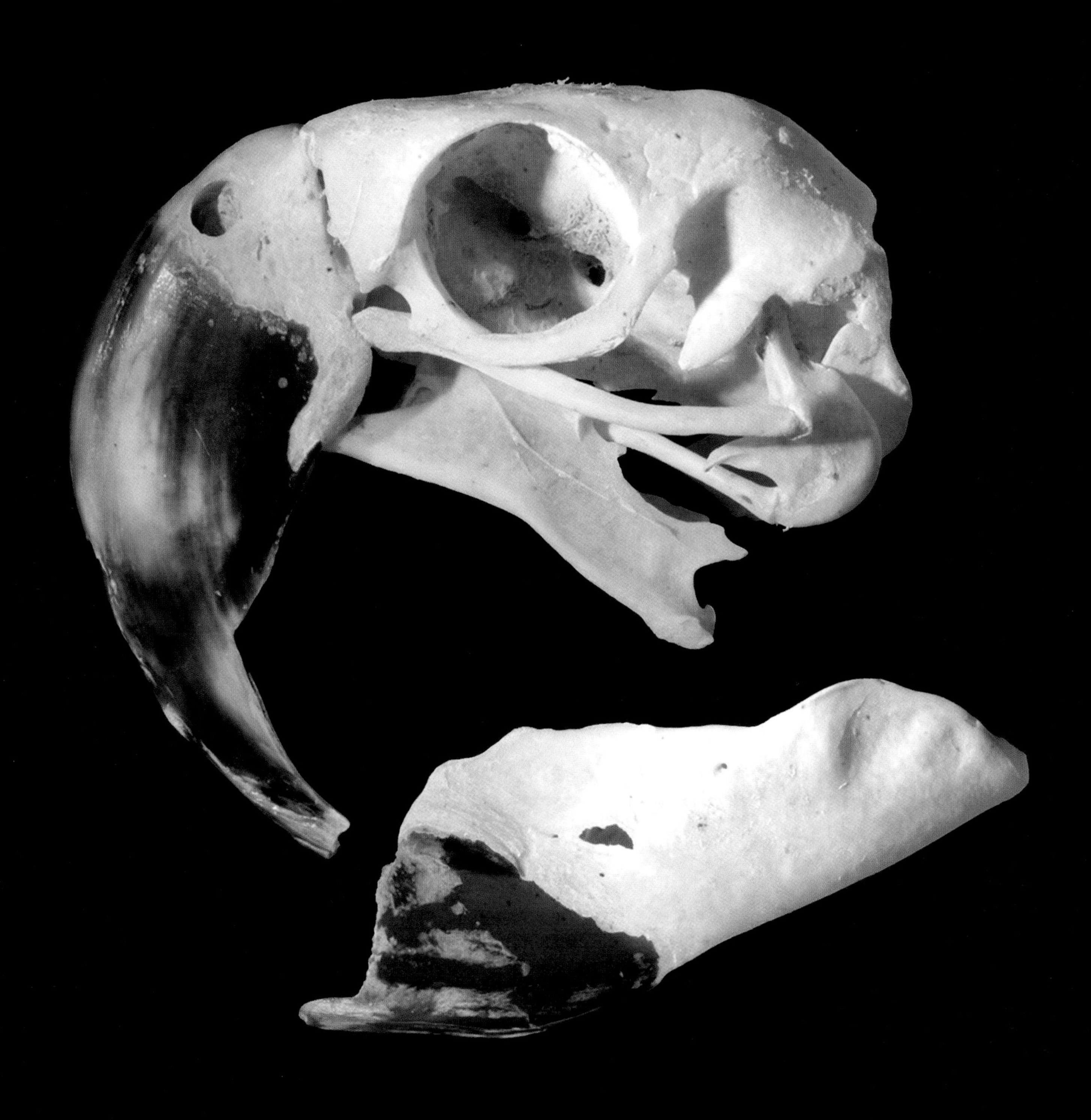

PLATE 30

PLATE 31

BALANUS (TAMIOSOMA) GREGARIUS, **A BARNACLE.**

Pliocene (*ca.* 3 million years).
Salinas Valley, California.

This giant barnacle is elongated in a coral-like fashion—and is adhered into a small colony as well.

EXAPTATION: REFINING THE CONCEPT OF ADAPTATION

Wings of pterosaurs, birds and bats all evolved, independently, and somewhat differently, as adaptations for flight. They were, in each instance, modifications of the forelimbs of walking tetrapod animals. But those ancestral, walking forelegs were themselves originally adaptations. In a sense, evolution is a matter of co-opting pre-existing adaptations—pieces of anatomy and behaviors that evolved under natural selection to perform specific functions—and transforming them into new structures and functions that serve different purposes. Paleontologist Elisabeth Vrba (with Stephen Jay Gould; the concept was originally Vrba's) coined the term "exaptation" for the phenomenon of adaptive structures becoming co-opted for different uses.

Her original example was as simple as it was elegant: the African black heron uses its ordinary heron-like wings to fly just as all herons do. But in this species, black herons have evolved the behavior of folding their wings into a sort of umbrella, which casts a dark, circular shadow on the water. The heron keeps its eyes on the water under its wings, its sharp beak ready to grab the fish that are attracted to this circle of darkness on an otherwise bright African day. The wings have become a sort of trap; and, for the moment, transformed from a flying adaptation to an adaptation for feeding.

Vrba's concept of exaptation takes away some of the surprise when we learned, rather recently, that many dinosaurs, including the mighty tyrannosaurs, themselves were covered with feathers. [Plate 1, see page 2.] Feathers are thermoregulating devices, helping to maintain body temperatures needed by active predators. Mere possession of feathers by no means implies that the animal can fly. But feathers, it turns out, were also precursors for flight: birds are feathered dinosaurs that developed the capacity to fly—through a series of intermediate steps probably at first involving simple gliding. The forelimbs were modified eventually into structures that were no longer any good for scrambling over the ground, but were instead co-opted into wings. And the feathers, still acting as thermoregulatory organs, were co-opted as essential elements of bird-style winged flight.

In other words, everything comes from something.

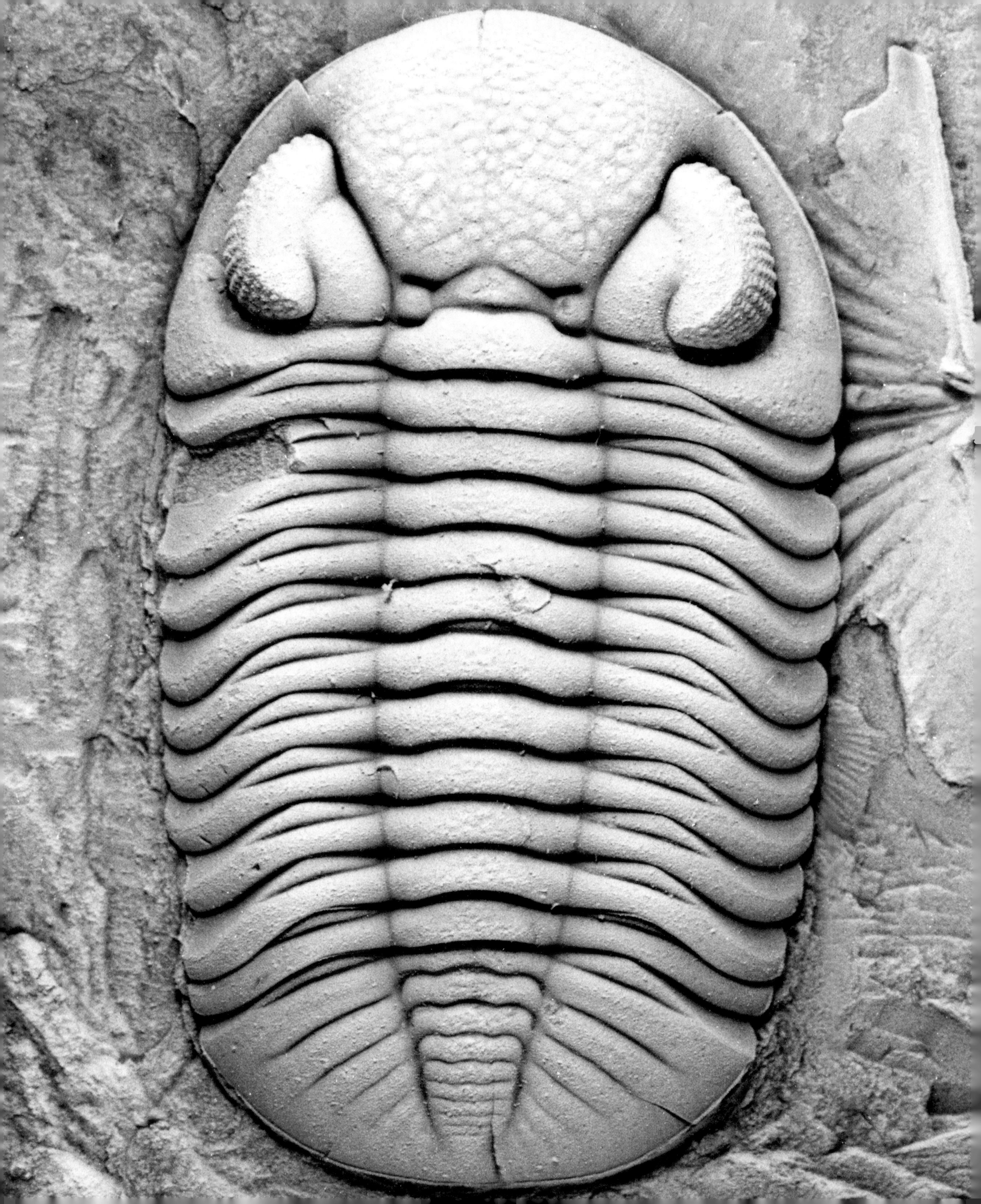

CHAPTER 3

ORIGIN OF SPECIES

PLATE 32

PHACOPS MILLERI,
A TRILOBITE.

Middle Devonian (*ca.* 380 million years). Silica Shale, Sylvania, Ohio.

Phacops milleri is one of the oldest species of the *Phacops rana* lineage. This species has eighteen vertical columns of lenses ("dorsoventral files") in each eye of well-preserved mature specimens, such as the trilobite shown here (along with the iconic Middle Devonian brachiopod *Mucrospirifer*). It was through careful study of the eyes of these and related trilobites, noting the distributions of the lens counts in time and through space, that first led me to the concept of "punctuated equilibria."

Credit: G. Robert Adlington/ Niles Eldredge.

Imagine setting out on an expedition to collect a large sample of fossils. Your goal is to document evolutionary change through geological time. You have done your homework: you have selected a type of fossil that seems perfectly to fit the bill as an ideal evolutionary subject. It is a species that is known to have existed for at least 6 million years. It is a fairly common species and occurs throughout a wide range of environments spread out over half a continent. Specimens are not hard to come by, and they usually are quite well preserved. Best of all, these fossils display discernible complex anatomical features, increasing the chances that you will be able to detect some evolutionary change as you trace these creatures' collective history up through their record in time. You have found the ideal evolutionary experiment, already performed and frozen in the rocks, just waiting for your hammer and chisel, your laboratory analysis, and, ultimately, your reflective contemplation. [Plate 33]

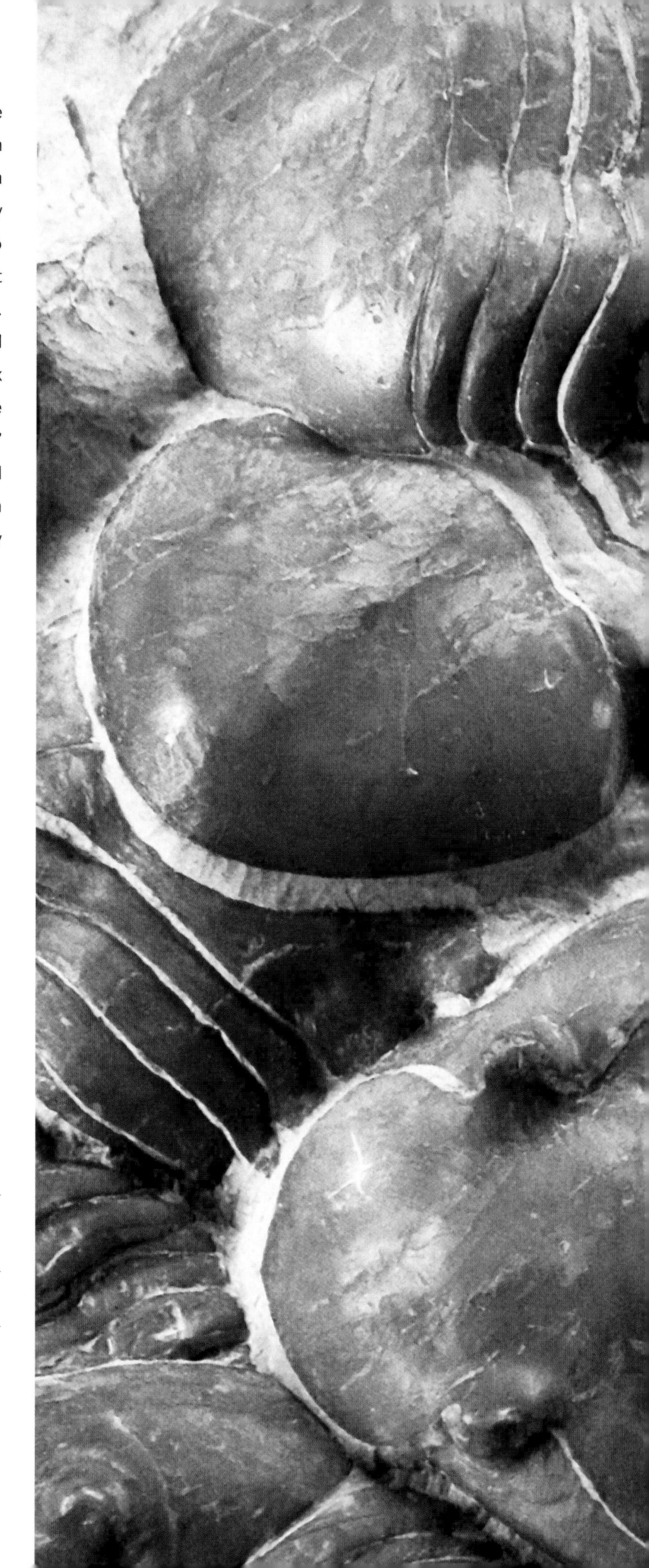

PLATE 33

***HOMOTELUS BROMIDENSIS,* SLAB WITH MANY SPECIMENS OF THIS TRILOBITE SPECIES.**

Middle Ordovician (*ca.* 470 million years). Criner Hills, Oklahoma.

Richly fossiliferous slabs such as this offer paleontologists the chance to study variation and evolution in extinct species.

PLATE 34

***RHYSONETRON BYEI,* THOUGHT BY SOME PALEONTOLOGISTS TO BE THE TRACES OF THE OLDEST KNOWN FORM OF COMPLEX ANIMAL LIFE.**

Precambrian (*ca.* 2 billion years). Flack Lake, Ontario, Canada.

Sometimes fossils consist solely of marks left from the activities of organisms.

It is over 150 years since Darwin's *On the Origin of Species* has appeared. In it he expressed concern that the fossil record had not as yet fully confirmed his thesis that evolution is for the most part a matter of smooth, steady, gradual progressive modification through time. He blamed paleontology's infancy for the lack of examples of what he called "insensibly graded series" through the rock record. Few regions had yet been explored for their paleontological treasures by the mid 1800s. And he pointed to the vagaries of fossilization as perhaps the main reason why we might forever be doomed not to find the kind of evolutionary sequences that Darwin felt his theory predicted—or even demanded.

Darwin knew that much time is actually missing in the sedimentary deposits that have turned to fossiliferous rock over the ages. He knew, too, that most organisms disappear without a trace upon death: only a very few escape the ravages of total decomposition and dissolution, leaving their hard parts as traces of their former existence interleaved with the layers of mud, silt and lime that form sedimentary rocks. [Plates 34, 35] Fewer still survive the vicissitudes of geological change, where whole mountain systems are eroded to low stubs, destroying all the contents of their rocks. Or where sediments are squeezed and baked in a metamorphic spasm that obliterates their fossil content. Then, of course, there is also the great possibility that fossils will forever lie buried beneath thousands of feet of younger rock, never to be exposed to human view. And even if fossils are exposed on Earth's surface, what are the chances they will be spotted by a knowing eye and brought back to the lab before they fall apart in a few seasons of exposure to the elements?

PLATE 35

***HIPPARIONYX PROXIMUS,* MOLD OF VALVE OF BRACHIOPOD SHELL.**

Lower Devonian (*ca.* 390 million years). Knox, New York.

In this specimen, the calcareous shell dissolved away, leaving a hardened mud imprint of the interior anatomy which is nonetheless full of anatomical detail.

Darwin thought he had explained why there were no really convincing examples of gradual evolutionary change in the annals of paleontology as he was writing. But he also saddled himself with the prophecy that his theory would ultimately stand or fall on the successful demonstration that the fossil record does indeed yield examples of slow, steady, progressive evolutionary change.

Now, in the twenty-first century, it is time once again to tackle the Darwinian challenge to paleontology: document in the fossil record evolutionary change as an inevitable and inexorable adaptive transformation through time. That is what you have resolved to do—with heightened technology, much known about geology, and all the advances in laboratory analysis that should make such a study far easier to carry out than in Darwin's day. So you take a few years, collecting one kind of trilobite all over the eastern and central United States. You collect in the muddy nearshore environments of the east. You sample the limestones and limey muds of the Midwest.

The equator had run through North America back then in the Devonian Period [Plate 36], and you find yourself collecting these trilobites in and around small fossilized coral reefs. Life fairly teemed in those Devonian tropical seas, with hundreds of species of invertebrates—brachiopods, clams, snails, nautiloid and ammonoid cephalopods, corals, bryozoans and crustaceans in addition to the beloved trilobites—present in profusion. A fair number of them left their hard outer shells behind in the muds. Middle Devonian rocks of North America, as in many other places throughout the world, are often so rich that fossils take up more space than the muds and silts that cement them all together. This fauna lived, through shifting sea levels and alterations of kinds of sea bottom for some 6 million years. Under the circumstances, confirming Darwin's vision should be a piece of cake.

PLATE 36

BOTHRIOLEPIS CANADENSIS,
A PRIMITIVE JAWLESS FISH.

Upper Devonian (*ca.* 350 million years). Gaspé Peninsula, Canada.

These extinct fish, of uncertain evolutionary relationships, thrived in the shallow seas of the Upper Devonian in North America. (See plate 128.)

PLATE 36

PLATE 37

PHACOPS SP.,* TRILOBITE CLOSELY RELATED TO AMERICAN *PHACOPS MILLERI* AND *PHACOPS RANA.

Middle Devonian (*ca.* 385 million years). Morocco.

Specimens from North Africa and Germany are virtually identical to species of the *Phacops rana* group found in slightly younger rocks in North America.

PLATE 38

***PHACOPS* SP., SIDE REAR VIEW OF SAME SPECIMEN AS IN PLATE 37.**

Europe and Africa collided with eastern North America approximately 38 million years ago; many species, such as this *Phacops*, invaded the inland seas of North America shortly after the continents collided.

PLATE 39

Imagine, then, your frustration when the suspicion begins to dawn, even before you get back to the lab, that all these trilobites look very much alike. [Plate 37] Regardless of where you are–say in the limestone quarries of Michigan, or the silty outcrops in the folded hills of central Pennsylvania, one trilobite looks much like another. You search the lowest (meaning the *oldest*) rocks you can find. You sample the *youngest* rocks, too, the very last vestiges of your trilobite before it disappears completely and forever from the ledger of time. And you begin to wonder, because even they, the very oldest and youngest samples, separated by 6 million years, look pretty much the same. [Plate 38] Far too much the same if the aim is to document the inevitability of evolutionary transformation through time. Maybe, you think, you just don't know these fossils well enough to see the differences, the evolutionary change that a more seasoned observer would detect right off the bat. [Plate 39] Maybe, when you get back to the lab, examine and measure all these specimens with your microscope, your computer and you will pick up some evolutionary change that your unaided eyes can't see. Maybe.

PLATE 39

***PHACOPS MILLERI*, TRILOBITE.**

Middle Devonian (*ca.* 380 million years). Sylvania, Ohio.

This species, and its close relative *P. crassituberculata*, lived unchanged for several million years in the clear seas that covered what is now the American midwest.

This is a true story, and it happened to me. Nor was my frustration purely intellectual: I had a Ph.D. thesis to write, and I needed *positive* results to show that I could formulate and carry out a scientific study on my own. It took awhile before I realized that the astonishing lack of change in these trilobites actually meant something–something rather important about the evolutionary process. And when I eventually did find some evolutionary change in my fossils, it came in a pattern that I hadn't expected: instead of progressive gradual transformation modifying the trilobites wholesale through time, I found that evolution seemed to be associated with little side-branches of new species splitting off from the main stock, and these descendant species living on side-by-side with the ancestral stock.

These, then, are the main ingredients of the theory of "punctuated equilibria": tremendous evolutionary conservatism, where little or no change tends to accumulate through time, and the concentration into branching events of what evolutionary transformation does occur. Stephen Jay Gould and I, in our 1972 paper that discussed these phenomena and coined the term "punctuated equilibria," dubbed the long periods of non-change "stasis." [Plate 40] It turns out that Darwin's contemporaries in the paleontological world knew full well that, once a species puts in an appearance in the fossil record, it tends not to exhibit much change throughout its stay up to its eventual disappearance. All of the paleontologists who wrote book reviews of the *On the Origin of Species*, in fact, commented that Darwin seemed to be ignoring this salient fact of the fossil record. And, though Darwin did briefly allude to this stability of species in his sixth edition of *The Origin of Species*, he basically stuck to his preference for a picture of gradual change. For over a century, the phenomenon of stasis was virtually swept under the rug. It was an ugly fact that seemed (to Darwin and many of his descendants) to threaten the very notion of evolution.

PLATE 40

***ANTHOSCYTINA PERPETUA,* COPULATING FROGHOPPERS.**

Middle Jurassic (*ca.* 165 million years). Jiulongshan Formation, Daohugou, People's Republic of China.

Behaviors—whether reproductive or not—are seldom so graphically preserved in the fossil record. This dual specimen is the basis for the conclusion of paleontologists that frog hopper anatomy and copulating behavior has remained unchanged for over 165 million years!

Credit: Dong Ren, Capital Normal University.

PUNCTUATED EQUILIBRIA: THE ALTERNATIVE TO GRADUALISM

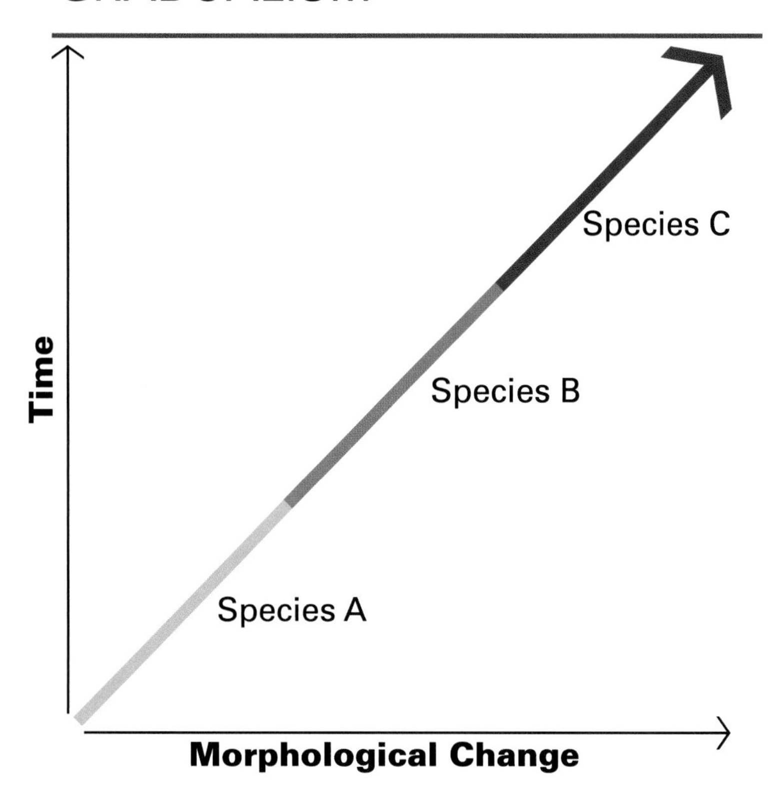

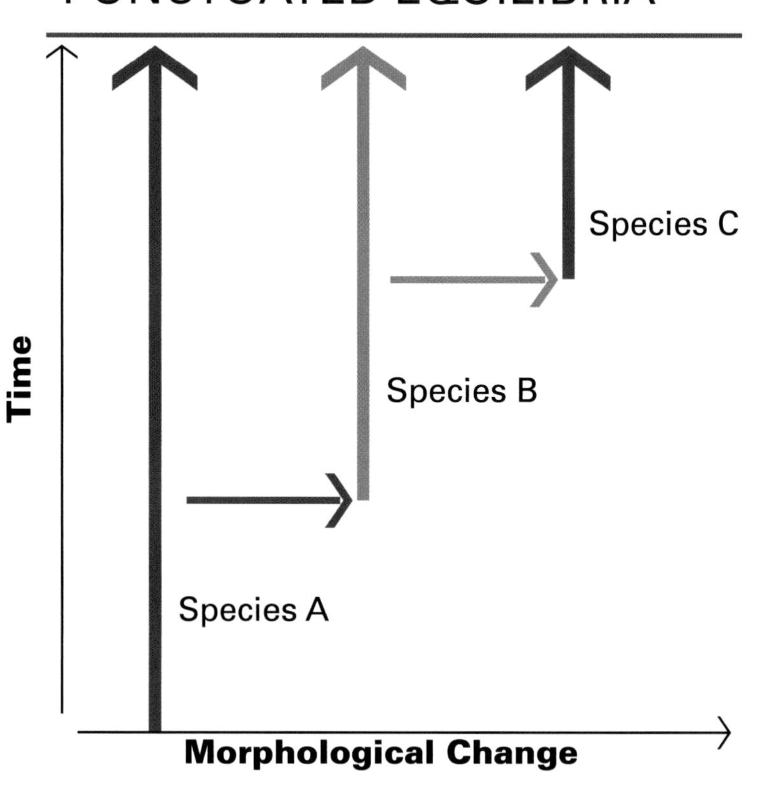

GRADUALISM VS. PUNCTUATED EQUILIBRIA

The difference between Darwin's theory that evolution happens gradually over very long periods of time (gradualism) and the theory of Punctuated Equilibria, which I developed is shown in these two diagrams. In Gradualism Species A slowly changes and adapts until it evolves into Species B and the process repeats itself until Species C evolves from Species B. In this theory, Species A and B only survive in the fossil record. In Punctuated Equilibria the evolution of Species B from Species A happens quite rapidly in geographic isolation, often in response to environmental change. In this model, Species A often survives alongside Species B.

On the other hand, the pattern of transformation occurring through branching seemed to me more comfortably familiar and rather easily explained. Much of the exciting work in evolutionary biology in the 1930s and 40s had dwelled upon the phenomenon of *speciation*: the splitting off of a new "daughter" from an ancestral species. Ernst Mayr, in the vanguard of evolutionary biology, remarked in 1942 that Darwin, ironically, had not discussed the origin of species in his *On the Origin of Species*. Mayr, along with the great geneticist Theodosius Dobzhansky, realized that adaptation through natural selection produces a continuous array of varying forms—what Darwin himself had called "insensibly graded series" of forms through time. But when we look at modern life, we notice that it comes in discontinuous, discrete "packages": there is little or no intergradation between these different groups we call "species."

Nor, these biologists realized, do the members of one species generally reproduce with members of another species. It was Dobzhansky's and Mayr's great contribution to recognize that new species arise from old when some organisms no longer interbreed with other organisms descended from the same ancestral species. Species are communities of organisms who interbreed with one another. When a new species splits away from its ancestor, barriers to reproduction arise between them. Darwin had furnished an explanation of adaptive diversity. Mayr and Dobzhansky told us why diversity is not smoothly continuous, but rather comes in those discrete packages we call "species." Darwin had skirted the issue—preferring to think that gaps between species alive today simply reflect long periods of evolutionary transformation after gradual divergence from a common ancestor. As to the fossil record, Darwin felt the gaps between species are simply the artifact of an imperfect record. As we shall see, the notion that species are real entities, reproductive communities, each with its own origin, history and termination (extinction) is a key to understanding many aspects of evolution—starting with punctuated equilibria.

Only recently have I discovered that Darwin had, as a young man in his 20s, also seen the outlines of punctuated equilibria. He saw that the birth of species in isolation (the idea of geographic speciation of Dobzhansky and Mayr, so essential to the notion of "Punctuated Equilibria") would account for the persistence of species, unchanged, "through thick formations"—in other words, our concept of "stasis." Darwin contrasted this vision with the inevitable gradual change of species—a vision of evolution he came to favor and promote, though he lacked empirical evidence for it.

With the birth of species in isolation, Darwin reckoned that adaptive change through natural selection happens rapidly in small populations. But with the passage of geological time and the inevitable environmental change that occurs, Darwin thought that natural selection would be constantly modifying entire species slowly and gradually. He could not reconcile the two views—and so his problem was deciding which was the most likely context for adaptation via natural selection to occur. He chose gradualism.

The theory of punctuated equilibria confronts the phenomenon of stasis squarely. The long period of stasis that characterizes the vast bulk of the history of most species constitutes an "equilibrial" condition. The "punctuation" comes in when a new daughter species buds off from the original parent species, a process that takes thousands of years, rather than the much longer period of stasis, which typically consumes millions of years. Events that take even a few tens of thousands of years can be very difficult to detect in the fossil record. Thus punctuated equilibria offers an additional explanation for why there are usually gaps even between closely related species in the fossil record: the transition from parent to daughter species takes place too quickly, and in such a limited part of the parental species' range, that there is little chance that the event will be recorded at all in the fossil record. [Plate 41] What we tend to see in the fossil record are successful, long-lived, far-flung species—species that remain, in the vast majority of cases, very much recognizably the same throughout their entire existence. [Plate 42] *Stasis is the single most compelling fact of the fossil record*, and one deserving further attention before we return to the task of understanding how evolutionary transformation does, in fact, happen. [Plate 43]

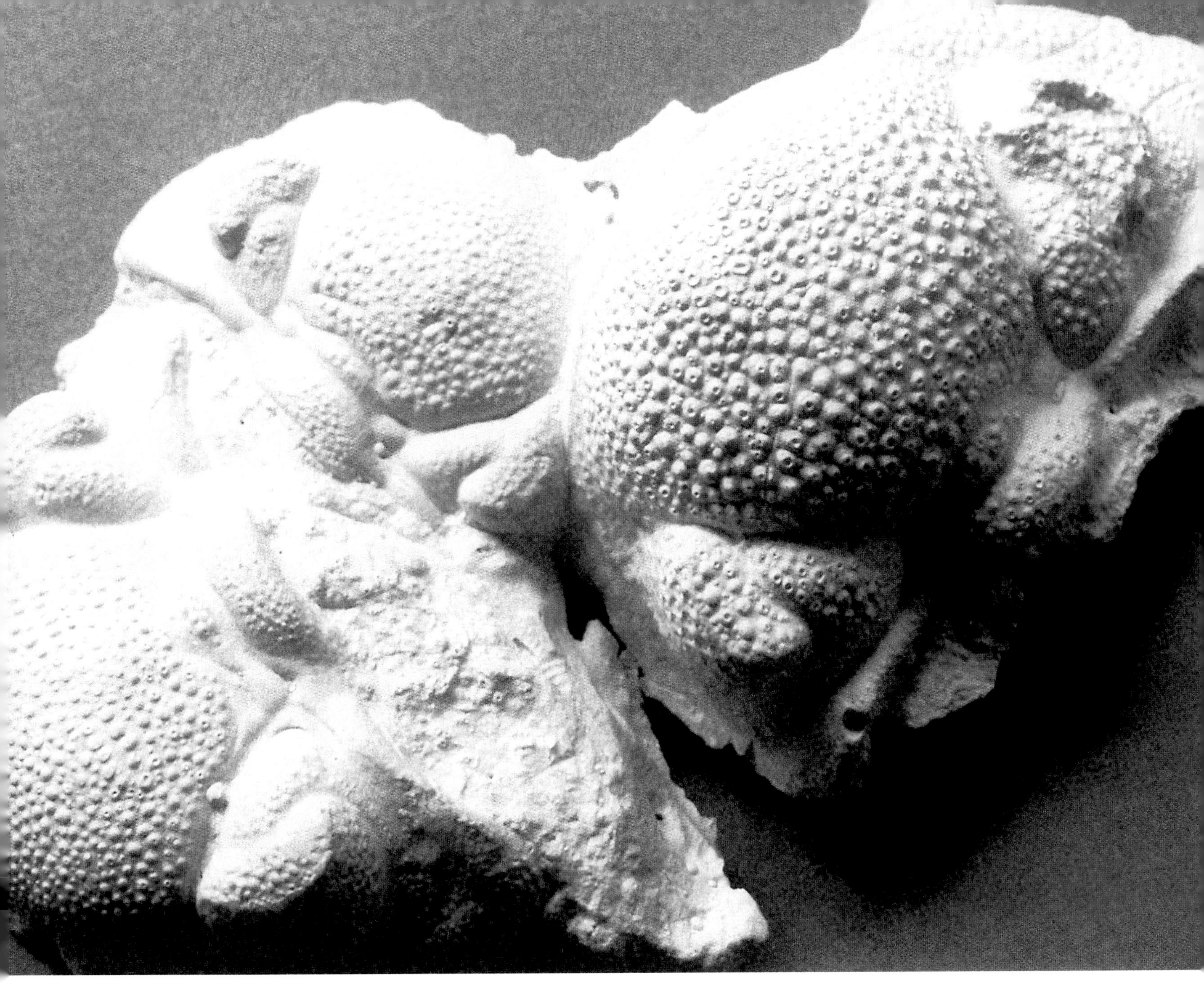

PLATE 41

***PHACOPS RANA* (LEFT TWO) AND *PHACOPS IOWENSIS* (RIGHT), CAST OF SLAB WITH THREE TRILOBITE HEADS.**

Middle Devonian (*ca.* 385 million years). Arkona, Ontario, Canada.

These two species are rarely found together; both species appear to have invaded the Arkona region only after *Phacops milleri* had become extinct.

PLATE 42

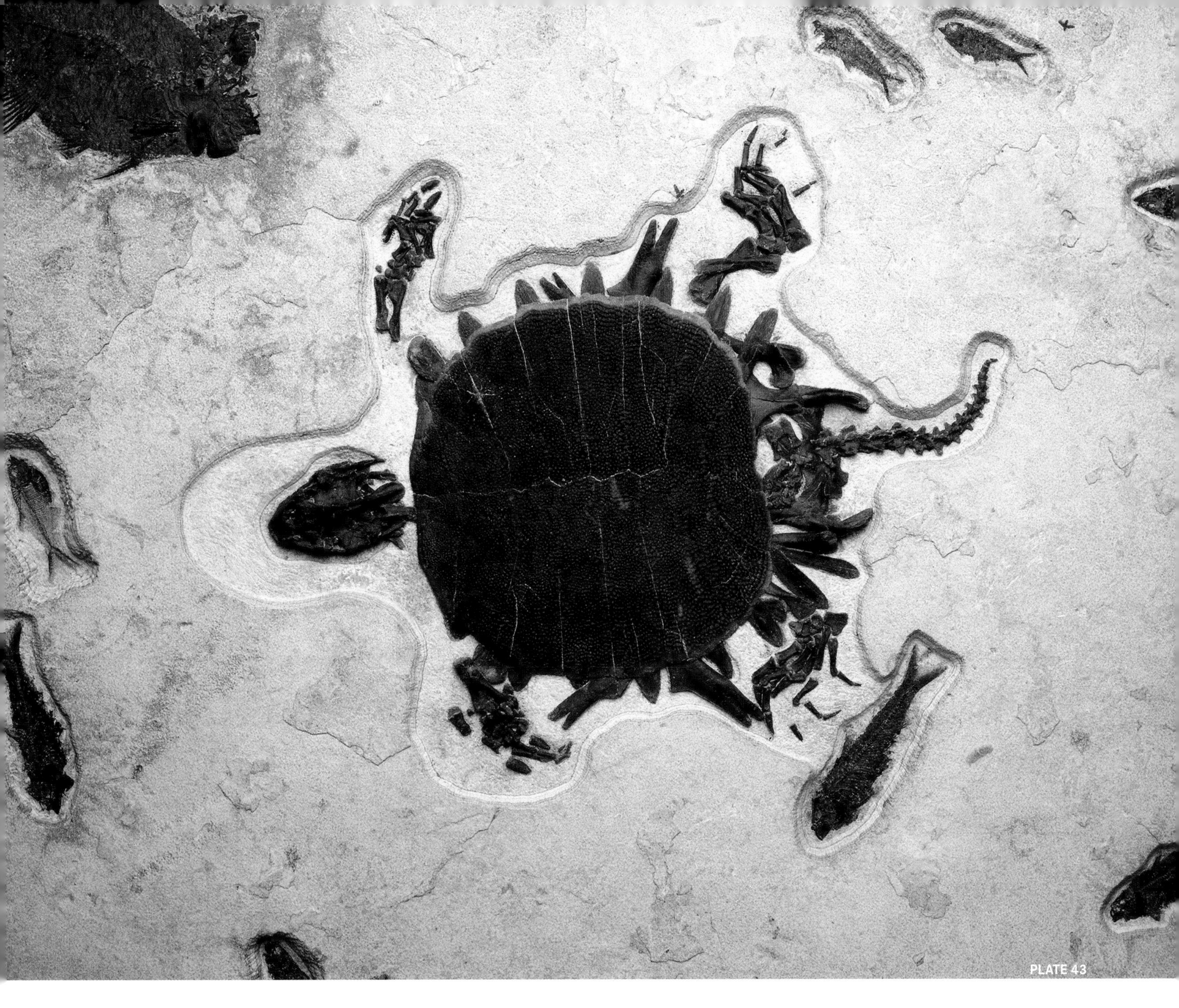

PLATE 42

***VIVIPARA CONTECTOIDES,* A LAND SNAIL.**

Modern specimens have been attached artificially to the surface of the rock containing virtually identical Pleistocene fossils (*ca.* 1 million years). Bahama Islands.

The modern specimens were glued to the fossils to demonstrate the lack of change in this species through a million years of geological time.

PLATE 43

***APALONE HETEROGLYPTA* , A SOFTSHELL TURTLE.**

Lower Eocene (*ca.* 50 million years). Fossil Lake, Green River Formation, Wyoming.

Turtles first show up in the fossil record in the Upper Triassic about 220 million years ago. The anatomy of these earliest turtles, as well as this Eocene turtle, is remarkably similar to that of contemporary turtles.

Credit: Lance Grande, The Field Museum.

STASIS

Everyone knows what robins and bald eagles look like. Baby robins and young eagles differ somewhat from the adult form that they'll grow into. Female robins differ somewhat from the males. But, in general, robins are robins, and bald eagles are bald eagles, all across North America. The same holds for the local species all over the world: not just birds, but trees and insects, worms and fish, shrubs and mammals.

On the other hand, we humans know that no two of us, even identical twins, are really exactly alike. We vary, one from another. And when we take a close look at robins and bald eagles—and all other species as well—we find that, though our first perception may well be that they all look alike, actually there are subtle differences between one robin and the next. Evolution, after all, depends on the existence of variation, and variation is as much a fact of nature as is stasis.

Yet we can buy bird guides and safely identify the birds in the air from their pictures in the book. Whatever variation there may be, it generally isn't so great that we cannot use one or two illustrations to be confident that we have made the right identification. The amazing thing is that we can also produce guide books to ancient life that encompass the flora and fauna that lived millions of years ago. A picture of the brachiopod *Mucrospirifer mucronatus* from the lower part of the middle Devonian rocks of New York and nearby states serves perfectly well to identify other specimens of this species that were living elsewhere at about the same time. [Plate 44] But it also serves just as well to identify specimens of *Mucrospirifer mucronatus* that lived several million years later. And that is remarkable. [Plate 45]

PLATE 44

***MUCROSPIRIFER THEDFORDENSIS*, A BRACHIOPOD.**

Middle Devonian (*ca*. 386 million years). Arkona, Ontario, Canada.

This species differs but slightly from *Mucrospirifer mucronatus,* the dominant, and very stable, species from the eastern seaway.

PLATE 44

PLATE 45

***MUCROSPIRIFER MUCRONATUS*, A BRACHIOPOD.**

Middle Devonian (*ca*. 395 million years). Bolivia.

Mucrospirifer mucronatus is an iconic Middle Devonian brachiopod species from eastern North America. It varies but little in space and time—and is thus as easy to identify as is the equally iconic modern American Bald Eagle.

Credit: Roger Weller, Cochise College

There seems to be a conflict between the variation that occurs naturally within species and the phenomenon of stasis. Stasis implies that the variation of the moment is not routinely and inevitably translated into progressive change of entire species through time. But stasis does not imply that organisms do not vary: rather, that variation within species always occurs within certain limits. Some species are more variable than others (and hence require more pictures to be usefully depicted in a guidebook). We see variation within fossil samples: even in a groups of fossils collected from the same bedding plane (meaning they almost undoubtedly lived together at the same time), no two are exactly alike. And just as eagles vary a bit over the North American continent, we also see geographic variation in fossils. In my trilobites—the species *Phacops rana*, populations in the western seas of New York tended to have more lenses in their compound eye than did their same-sized counterparts living in the muddier environments further to the east. Through time, the average number of lenses in the eyes of all the samples of *Phacops rana* I had measured seemed to increase progressively through time. There are examples of variation in space and in time—and temporal variation can lead to net changes through time: good old, predicted Darwinian gradual evolutionary change.

But, as often as not, such change reverses direction. Those Pennsylvanian snails *Ananias welleri* and *Worthenia tabulata*, which (as we saw in Chapter 2) together show such striking convergent similarity, both changed through time. But their evolutionary course was zig-zag. Neither moved very far away from the shape they started out with. And that is the real point of stasis: it is not a rigidly total lack of variation or accumulated change through time. It is, rather, a wobbling around some initial state. Rarely does wholesale, directional change of an entire species ever seem to *get* anywhere; rather, while one species is existing, with some change here and there, over many millions of years, new species are arising, and others are becoming extinct. What has become apparent is that the major changes in evolution are tied up with the pattern of speciations and extinctions—rather than the generally insignificant modification that accumulates within a species once it has come into existence in the first place.

Does the reality of stasis really spell the death knell of the Darwinian view? Hardly, if by Darwinism we simply mean the very notion of evolution through natural selection. The Darwinian view, recall, sees natural selection charting nearly constant changes in environmental conditions. We have seen that there is indeed variation in nature; and there is no doubt that conditions at any one place seldom remain exactly the same for long. On the long-term scale, continents move around, are alternately high and dry or awash with shallow seas, and are sometimes subjected to waxing and waning glaciers. On the short term, three years of normal rainfall may easily be followed by two years of drought. At the intermediate level—one involving thousands, but not millions of years, perhaps the critical level in terms of evolutionary effects—we see major climatic changes happening all the time. Ten thousand years ago, Baja California was pretty green, and what are now New York and Moscow were covered with ice sheets many hundreds of feet thick.

Why, then, don't we see more examples of gradual evolutionary transformation tracking these real and progressive changes in the environment? How do we explain this general failure to observe the

PLATE 46

ECPHORA QUADRICOSTATA,
A GASTROPOD MOLLUSK,
OR SNAIL.

Upper Miocene (*ca.* 8 million years).
St. Mary's River, Maryland.

This species, the first from North America to be described in the scientific literature, fell prey to extinction when it could no longer find suitable habitat.

predicted match-up between environmental change and evolutionary response? The answer, in retrospect, seems simple. There is nothing whatever wrong with the basic Darwinian paradigm. Rather, the fault lies simply in what we imagine are the consequences to life when environments indeed change.

We know, of course, that some birds migrate south and some mammals hibernate, when winter comes. Retirees and others wealthy enough to follow suit head for warmer climes as well. The rest of us have adaptations to cope with colder conditions: hair grows longer and thicker (and perhaps white to match the snow) in non-sleeping mammals. And some of us turn up the thermostat. The point being that organisms, in one way or another, are constantly matching their physiologies with the ambient environment.

And that is precisely what happens during the longer-term (multi-thousands-of years) basically gradual changes in the physical environment. As climates change, as glaciers creep southward and mean annual global temperatures begin to drop, habitats are radically altered and *rearranged*. Plants that cannot bear subfreezing temperatures travel south. How does a plant move south? Obviously, an individual plant cannot itself get up and move, but species can. Seeds or spores, a plant's propagules, can and do indeed become cast far and wide, and those that happen to be transported further south will be the ones that take root, survive, and go on to produce further generations. And, in much the same way, the animals that depend on those plants end up where the plants have gone. The whole response is to continue to find *suitable habitat*—environmental conditions to which organisms are, through their past evolutionary histories, already accustomed.

The old Darwinian projection more or less assumed that, as environments change, organisms will, in effect, sit there, grin and bear it; if there is suitable variation for selection to work on, over the generations species will become adapted to the newly instituted conditions. What really happens most of the time, though, is that organisms seek out conditions to which they are best suited, abandoning habitats no longer to their (evolutionarily imbued) liking. So this process of habitat tracking is actually a form of natural selection! But in this case, the organisms doing the surviving and reproducing tend to be the ones that have found the recognizable living conditions to which they are already adapted. Natural selection is for the *status quo*. Even in the face of profound and prolonged environmental change, species will not change much as long as their members can find recognizably suitable living conditions. The alternative is extinction, which is far more likely to occur than progressive modification in response to new environmental challenges. [Plates 46, 47]

PLATE 47

***BARBOUROFELIS FRICKI,* A "PALEOFELID," CLOSE RELATIVE OF TRUE CATS.**

Upper Miocene (*ca.* 6 million years). Nebraska.

Organisms with specialized adaptations, such as this saber tooth with its exaggerated canines, are more prone to extinction than their more generalized relatives.

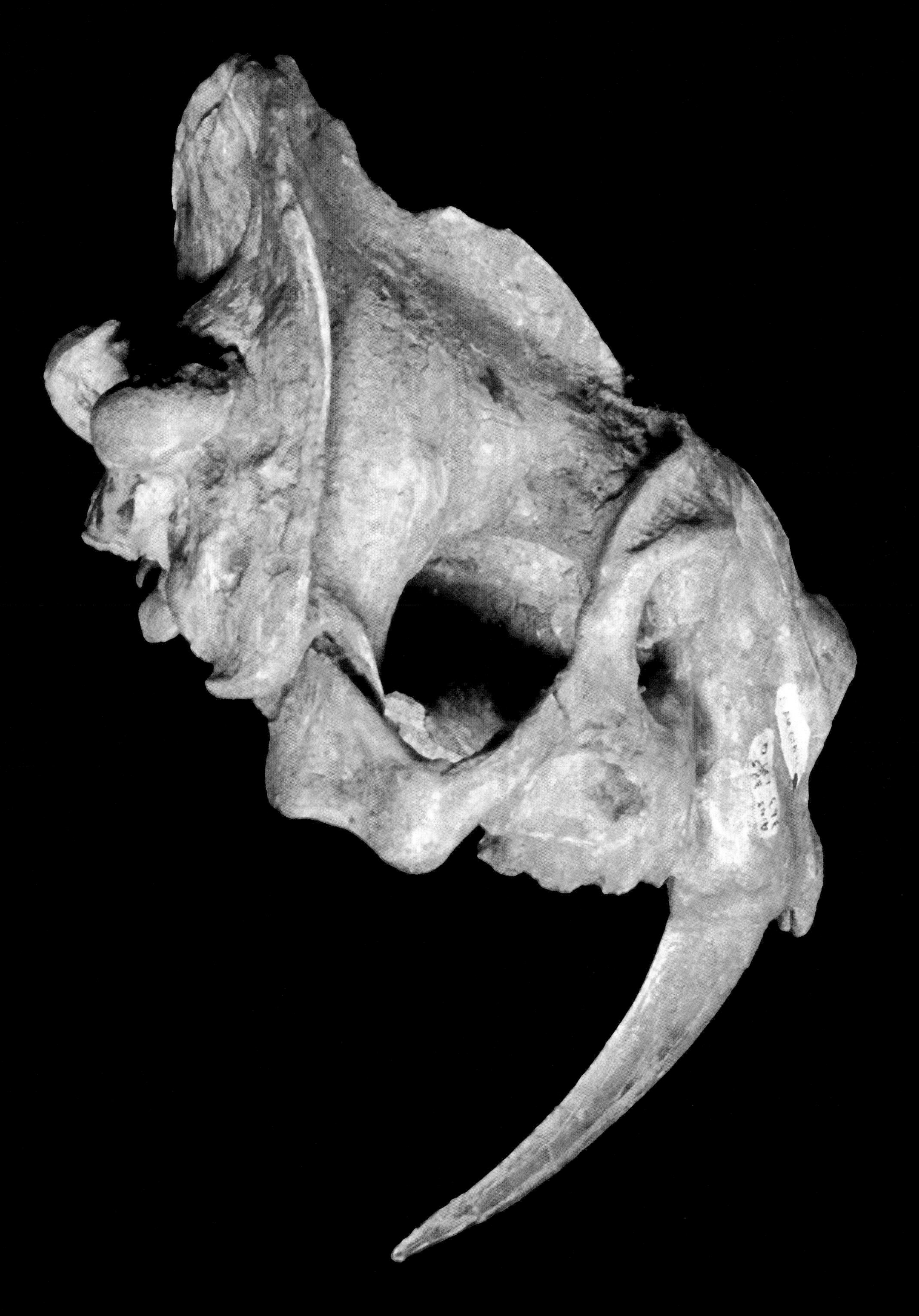

PLATE 47

PLATE 48

***PHACOPS* SP., CLOSE-UP OF EYE OF TRILOBITE CLOSELY RELATED TO AMERICAN *PHACOPS MILLERI* AND *PHACOPS RANA*.**

Middle Devonian (*ca.* 388 million years). Morocco.

It was change in the structure of the eye of species of *Phacops* that best documented evolutionary change in the lineage extinction than their more generalized relatives.

Important as habitat tracking is, there is another way to think about the causes of stasis. Many species have large geographic ranges. Local populations find themselves in slightly different habitats–with somewhat different predators, food sources, diseases, water supplies and so forth. Each local population, faced with these variant conditions of existence, are slightly differently adapted. It becomes a tall order, then, for an entire species to change in any one particular evolutionary direction as time goes by–whether or not significant environmental change is taking place.

EVOLUTIONARY CHANGE AND THE ORIGIN OF SPECIES

My trilobites, the *Phacops rana* complex of the Middle Devonian, did turn out to show some evolutionary change, after all. Even though these trilobites basically looked the same from start to finish in their six million year sojourn, there was that slight trend to increased number of lenses in the eye–which shows that some gradual transformation can indeed occur within the history of a species. And there was something else, as well: a change that took a lot of looking to detect because it was so simple and minor a transformation. But it was a change that, nonetheless, tells us much about how real evolutionary transformation occurs and accumulates in the history of life.

Closely related species generally look a lot alike. They are different species because, despite their similarities, members of one group reproduce only with one another, and not with their closely similar relatives of another group–another "species." But, by and large, the differences between closely related species are not great. The differences in numbers of columns of lenses between my trilobite samples (differences which did, after all, show up back in the lab under the microscope) are slight, but nonetheless real. The evolutionary pattern is unmistakable: when change occurred, it seems to have occurred relatively rapidly. Then there was no further change for millions of years until, once again another rapid evolutionary event occurred.

My trilobites have large compound eyes of a special sort: unlike insect eyes, which often have hundreds of tiny lenses covered by a single "cornea," the eyes of a *Phacops* have at most one hundred and thirty lenses. Each lens, covered with its own cornea, is large enough to be seen by the naked human eye. It is a relatively simple matter to count the lenses on a trilobite eye–in spite of the fact that the creatures have been dead for nearly 400 million years! Because trilobites shed their skin to grow, each molted skeleton has the potential for fossilization. And that happy fact lets us see what baby trilobites looked like before they became adults.

It turns out that lenses (naturally) are added to the eye as the eye gets larger along with the rest of the trilobite's body. But they do so in an orderly fashion: in *Phacops rana* (meaning "frog *Phacops*," so named because of the frog-like look imparted by those very same huge, bulging eyes), lenses are arrayed in seventeen up-and-down columns. Babies have fewer–say, twelve or thirteen columns. But as they matured, the trilobites added lenses to the bottom of each column, and added a few columns. When they hit seventeen columns, they stopped. Though there is a bit of variation, seventeen is the "adult stable" configuration of columns of lenses in the eye of *Phacops rana*. [Plate 48]

But 'twas not always thus. The oldest *Phacops "rana"* yet discovered have, not seventeen, but eighteen, columns of lenses. Originally classified as a separate "subspecies," *Phacops rana milleri*, they really were a separate species. As trivial as the differences between "*milleri*" and true "*rana*" are (one column of lenses in the adult eye, after all, is not very much) nonetheless the differences are stable over both a wide geographic area and, even more significantly, over a long period of time.

The "*milleri*" form lived in the seas of central North America for a period of some 3–4 million years. It seems to be the primitive form, as it is virtually identical to trilobites living in slightly older seas in the Old World—meaning, in this case, present-day Germany and North Africa. *Phacops rana* proper lived in the seas of eastern North America, in the muddy environments of what is now New York State and in the basin which eventually became the Appalachian Mountains. The two different forms of *Phacops* lived side-by-side (but never, apparently, actually together) in eastern and central North America for about 3 million years.

If the eighteen-column form (*milleri*) came from the Old World, where did the seventeen-column true *rana* come from? From *milleri*. [Plate 32, see page 50.] But how?—the two seem to have been contemporaries for most of their history. The answer appears to lie in speciation: *Phacops rana* seems to have evolved from *Phacops milleri* through the isolation of a small segment of the ancestral *milleri* species in the east. Indeed, I collected a small sample of specimens which seem, in several ways, to be intermediate between the two. The eyes bulge with more lenses than later *rana*, and several specimens have an eighteenth column of lenses like *milleri*. The sample is either that of a very primitive *rana*, or a very advanced *milleri*—and seems to me to be a luckily-encountered window into the brief period of transition between ancestral and descendant species.

But, for the most part, the fossil record simply shows two very similar, but consistently different, trilobite species living in adjacent regions. Both species remain recognizably the same throughout the 3–4 million years of time they shared. But then there was an environmental event that took its toll: the central North American seas dried up almost totally, and *Phacops milleri* disappeared, apparently forever. When the seas came back to the continental interior, the original fauna was restored more or less in the fashion it had been. But, crucial to the punctuated equilibria story, the species that came back to populate the newly reinvigorated marine habitat was not *milleri*, but *rana*. It is as if *rana* could not invade the lush habitats of the midwestern seas until its sister (or, actually, parental) species had become extinct. When the seas once again washed over the interior, and with *milleri* gone the way of the ages, *rana* was able to "recognize" the coralline seas of the Midwest as suitable habitat—and moved right in!

The picture is critical. Collecting trilobites in Ohio, Michigan or southwestern Ontario today, you notice a "jump" from eighteen to seventeen columns-of-eye-lenses *Phacops*, all at the same apparent point in time. It would be tempting to think that evolution works by sudden jumps—in defiance of the Darwinian credo. Or, keeping Darwin firmly in mind, it would be tempting to speculate that evolution was gradual, but that time was missing in the rock record, just at the boundary between the occurrence of eighteen- and seventeen-column forms fossilized in the Midwest.

But both interpretations would be fundamentally wrong: the seventeen-column form did not evolve in the Midwest.

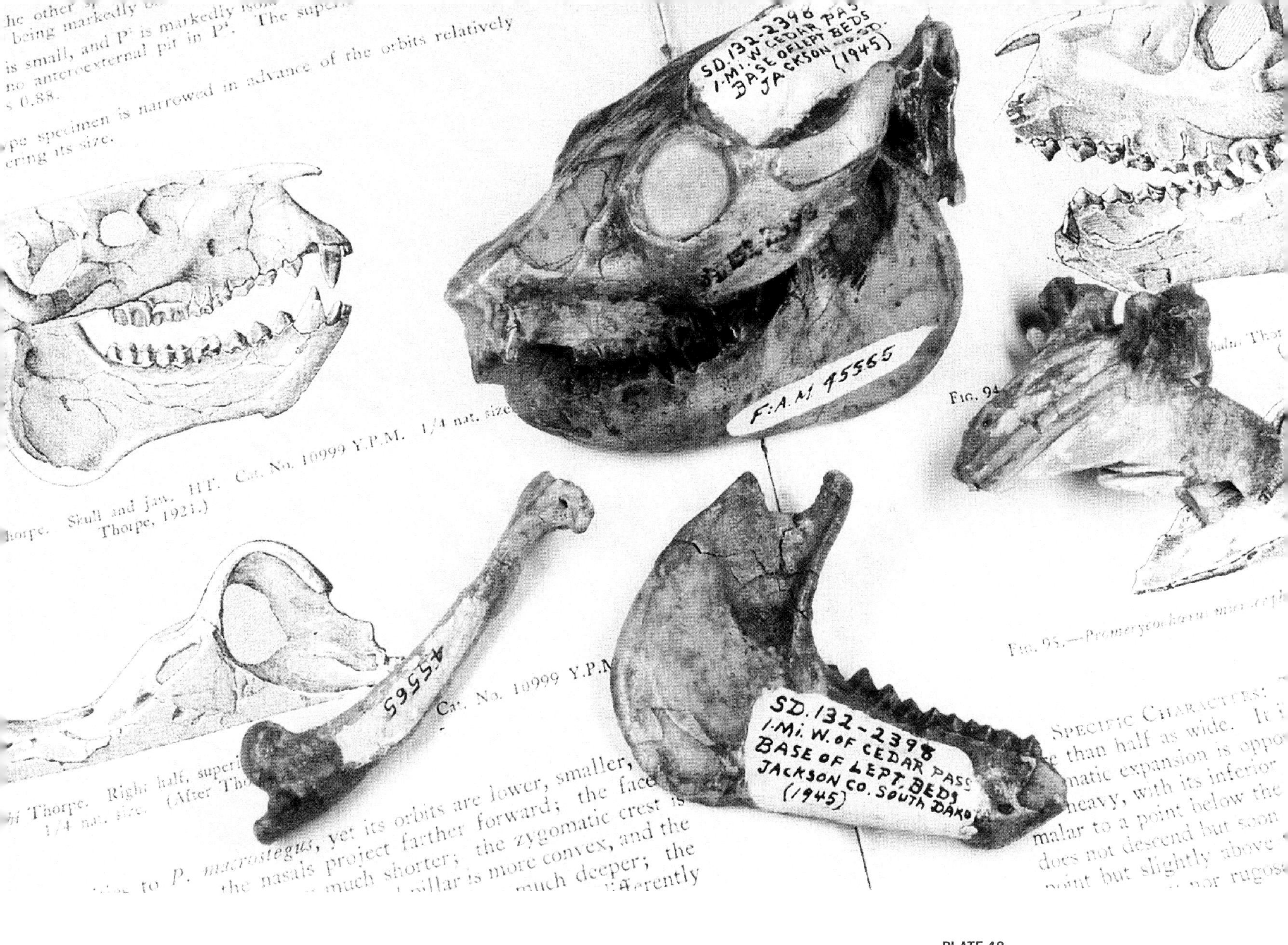

PLATE 49

***LEPTAUCHENIA* SP., A PRIMITIVE FORM OF ARTIODACTYL (CLOVEN-HOOFED) MAMMAL OREODONT RELATED TO HIPPOS AND PIGS.**

Oligocene (*ca.* 30 million years). Badlands of South Dakota.

Numbers of different, closely related species often coexist in the same ecosystem.

It had already evolved, millions of years earlier, in the East. Because closely related and ecologically similar species rarely can co-exist, I think the two species divided the territory, each occupying the habitats to which it was best suited. Each remained ensconced in its own bailiwick for some 3 million years. It was only after *Phacops milleri* had become extinct, after all, that *Phacops rana* proper was able to invade and live successfully in the midwestern maritime environments.

There are intermediates to be found in the fossil record—*if* you are a very lucky paleontologist. But they come packaged in a way not originally imagined by Darwin: rather than finding intermediate specimens temporally sandwiched between the older and younger samples, it is far more common to find no intermediates between an older, ancestral and younger, descendant species. Rather, it is more common to find the "descendant" already living while the ancestor lived—but in some other region. Just as parents in the vast majority of cases (thankfully) continue to live on after their children are born, new species arise from old and the two continue to coexist side-by-side. Evolution is as much a lateral (geographic) affair as it is a vertical, temporal one. New species arise from their "parents" by geographic isolation. A new species will persist only if it has a sufficiently different adaptation to the environment (a way of making a living distinct from its progenitor) that would allow it to survive alongside its ancestor.

So that, in its barest essential, is punctuated equilibria. The fossil record shows us that life does not evolve in a steady, progressive fashion. Most species actually remain pretty much the same throughout their existence. New species arise and, while they may not differ greatly from their ancestors, neither do they grade back into them as we trace their ancestry through time. Evolution is relatively abrupt—and I stress *relatively*, because Darwinian processes require but a few thousands of years (at most) to render the typical rather slight amount of change between a "parental" species and its descendant species. But thousands of years do not show up well in the fossil record, and we actually should expect that few "intermediates" will be found. This does not mean that evolution itself is totally abrupt, passing from one state to another without intermediate forms. But, especially back in the Paleozoic several hundreds of millions of years ago, the chances of finding the intermediates that lived so briefly over a few thousands of years are ever so slight, compared with the chances of finding the samples of the stable, far-flung ancestral species that might have lived as many as 5, 10 or even 15 million years—or the descendant species, which may well have ended up outliving its progenitor. [Plates 49, 50]

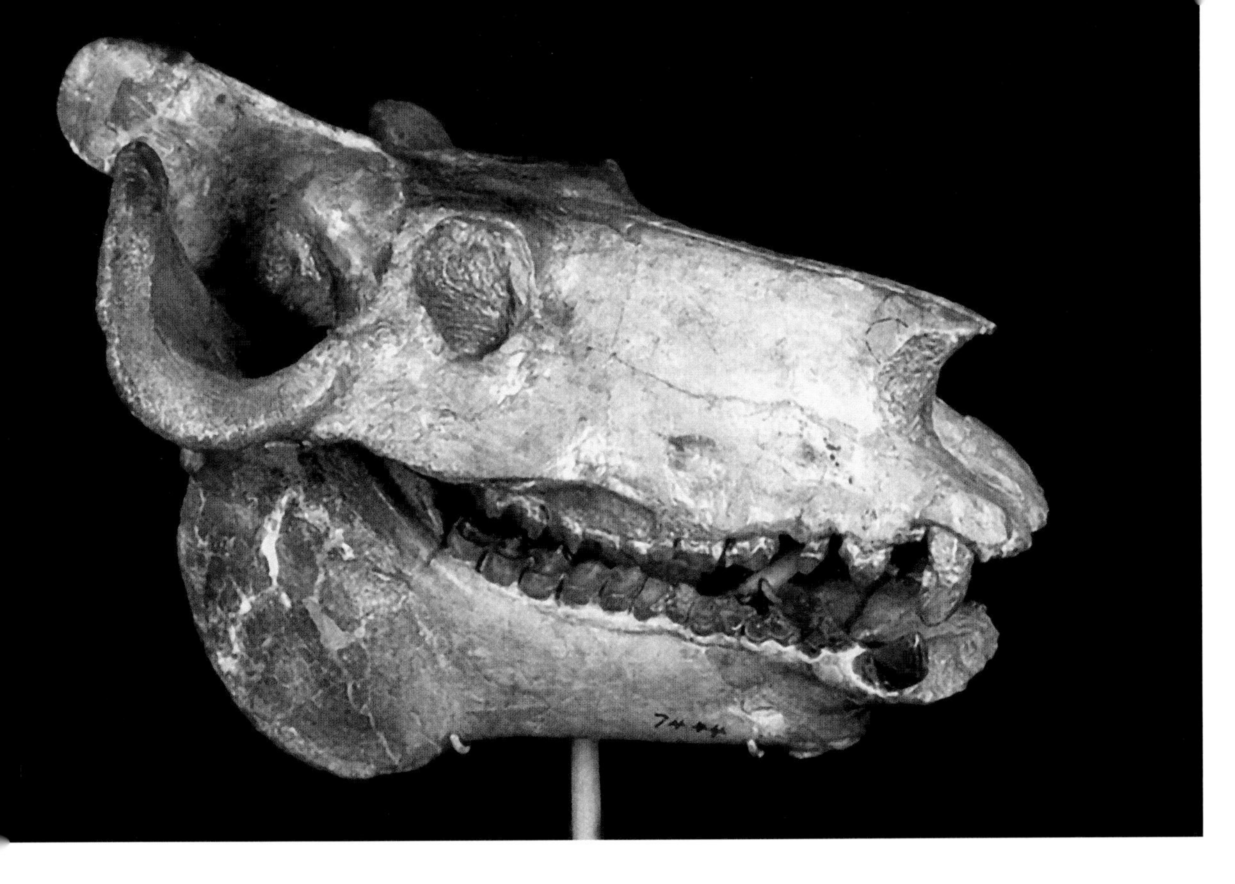

PLATE 50

***PROMERYCOIDON* SP., ANOTHER OREODONT.**

Oligocene (*ca.* 30 million years). South Dakota.

Oreodonts were very abundant in some Oligocene ecosystems of North America.

Evolution *looks* "jumpy" in the fossil record. It really is not: evolution really occurs essentially the way that Darwin said it does. But his imagined projection of what natural selection would produce over geological time, as well as the consequent patterns of stability and change that would show up in the fossil record, were a little off. Selection is going on all the time. But it is selection predominantly for constancy, for maintaining the status quo, so long as familiar habitats can be "recognized" somehow, somewhere, in the course of fickle environmental change through time.

Punctuated equilibria takes the real pattern of great stability that fairly leaps from the fossil record and puts it together, not with the old picture of gradual transformation of species, but with the newer picture of the emergence of new species through the fragmentation of lineages. It tells us that adaptive transformation does not occur to any significant degree unless speciation is involved. How this process works depends on other factors—particularly the ecological phenomenon of extinction. As we shall see in Chapter 7, large-scale evolutionary change (so-called "macro-evolution") depends very much on this same interplay between speciation and extinction. Through it all, the fundamental pattern of stability, "punctuated" by occasional fits of evolutionary change, rings clear as the most basic of all messages the fossil record has to tell us on the very nature of the history of life.

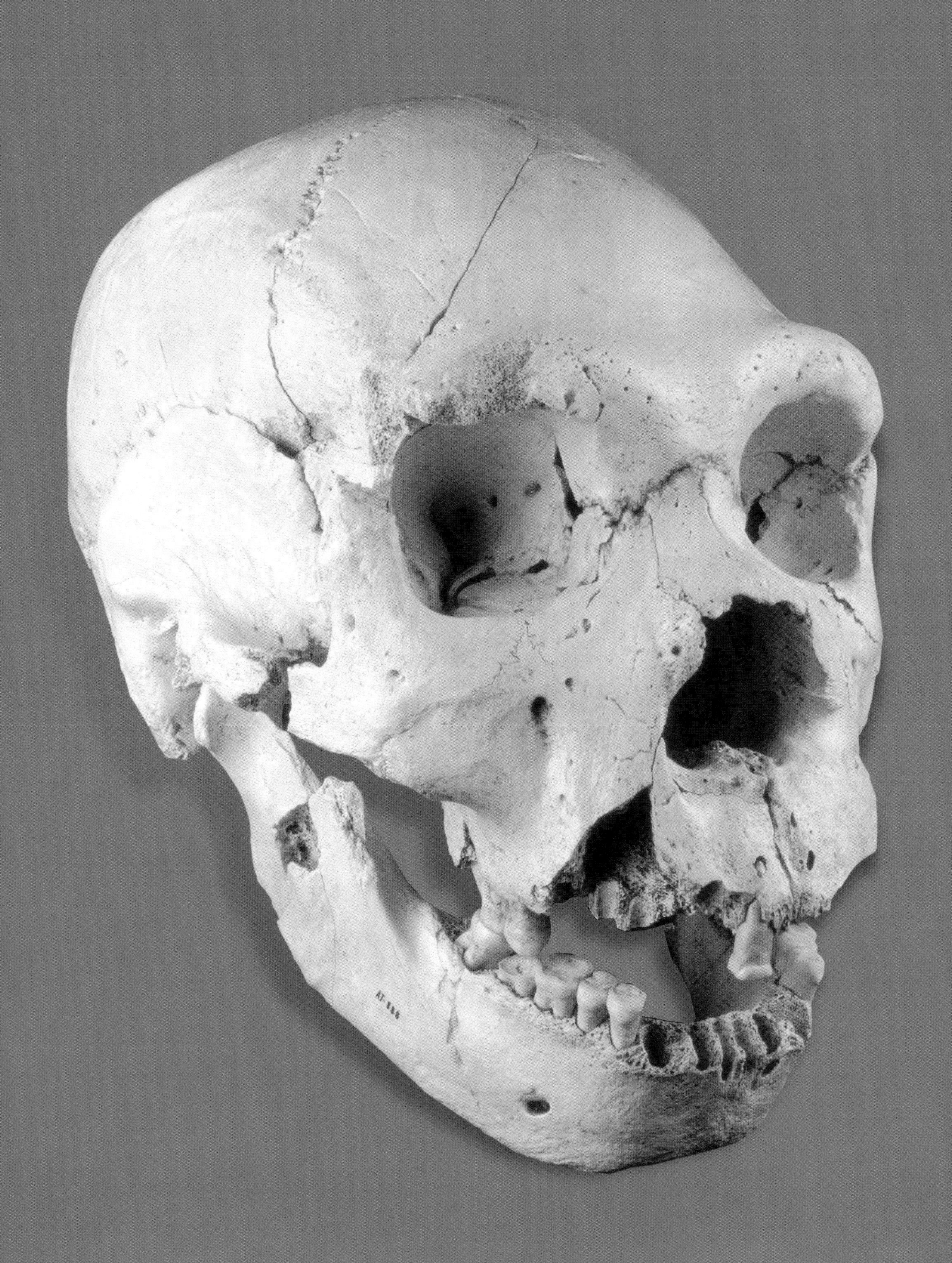

CHAPTER 4

HUMAN EVOLUTION

PLATE 51

HOMO HEIDELBERGENSIS (ANTECESSOR), **CRANEO 5.**

Middle Pliestoncene (*ca*. 350,000 years). Sima de lost Huesos (the pit of bones), Atapuerca, Spain.

More than 5,500 human bones representing twenty-eight skeletons have been excavated at the bottom of this 43-foot deep cave. Craneo 5 is the most intact skull recovered. Genetic evidence suggests these hominins are closer to the enigmatic Denisovans than to Neanderthals. (See Plate 66.)

Credit: Javier Trueba; courtesy Juan Luis Arsuaga, Universidad Complutense de Madrid.

PLATE 52

***HOMO SAPIENS,* MODERN HUMAN SKELETON.**

Recent.

The bones of a human skeleton match those of the skeletons of other mammals.

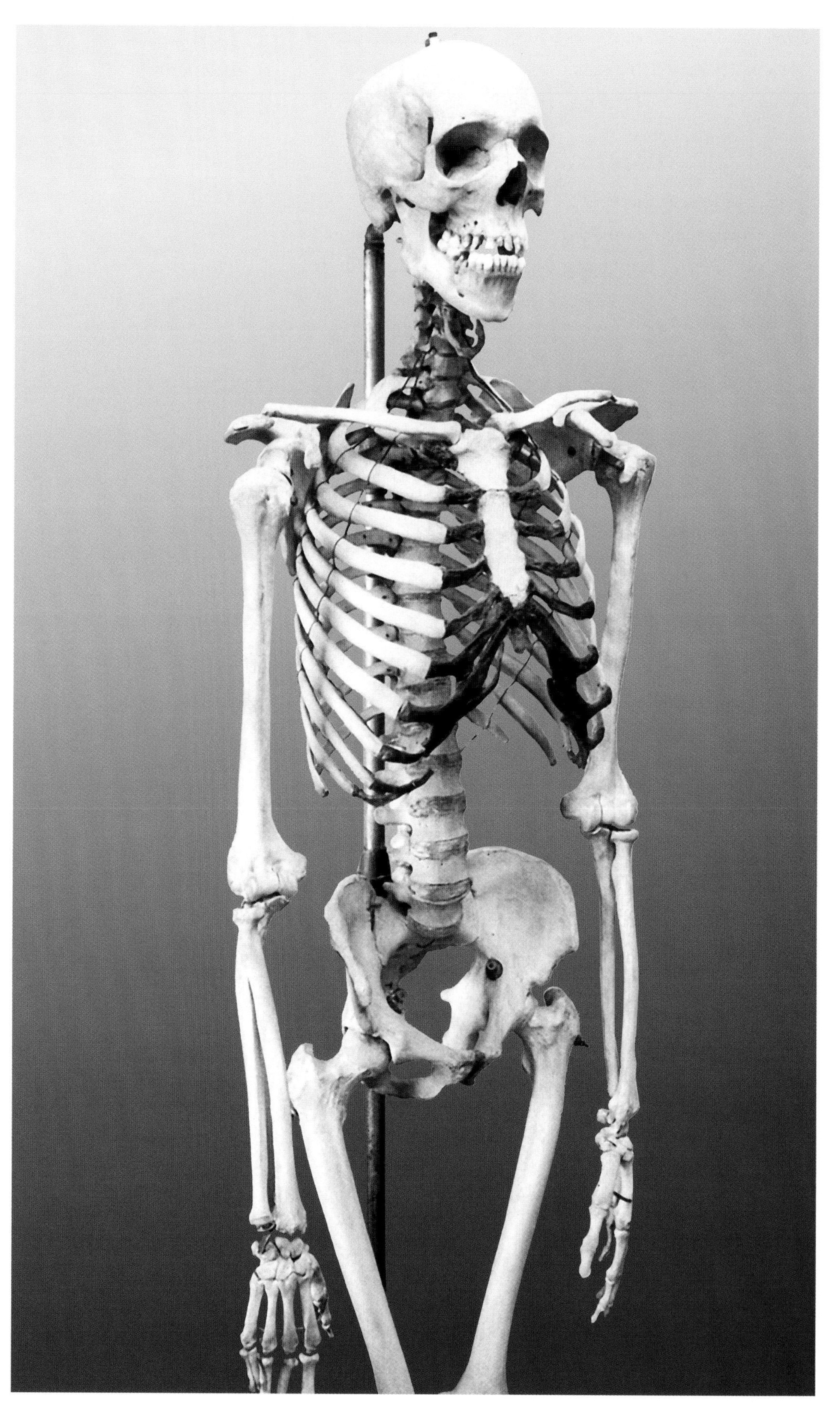

Darwin had electrified the thinking world. After 1859, all biologists saw the living world as a system with a past. The new activity of reconstructing life's past developed almost immediately, spearheaded by Ernst Haeckel, an influential German scientist who was also an accomplished artist. Evolution provided a wholly new way of looking at life in its myriad fossil and living forms.

Even when creationism held sway, and humans in particular were nearly universally held to have been created in God's image, it was nonetheless also understood that people are animals. [Plate 52] Domestic animals eat, sleep and reproduce in ways that are unmistakably akin to human behavior. The skeletons of cats and sheep are matched, bone for bone, in the human frame. [Plates 53] Once the idea took hold that the rest of the organic world is connected through a historical process of evolutionary ancestry and descent, it was inevitable that humans would be included in this new view of organic nature.

PLATE 53

SKELETON OF YOUNG FEMALE BABOON.

Recent.

The skeletal structure of non-human primates bears an especially close resemblance to the human frame.

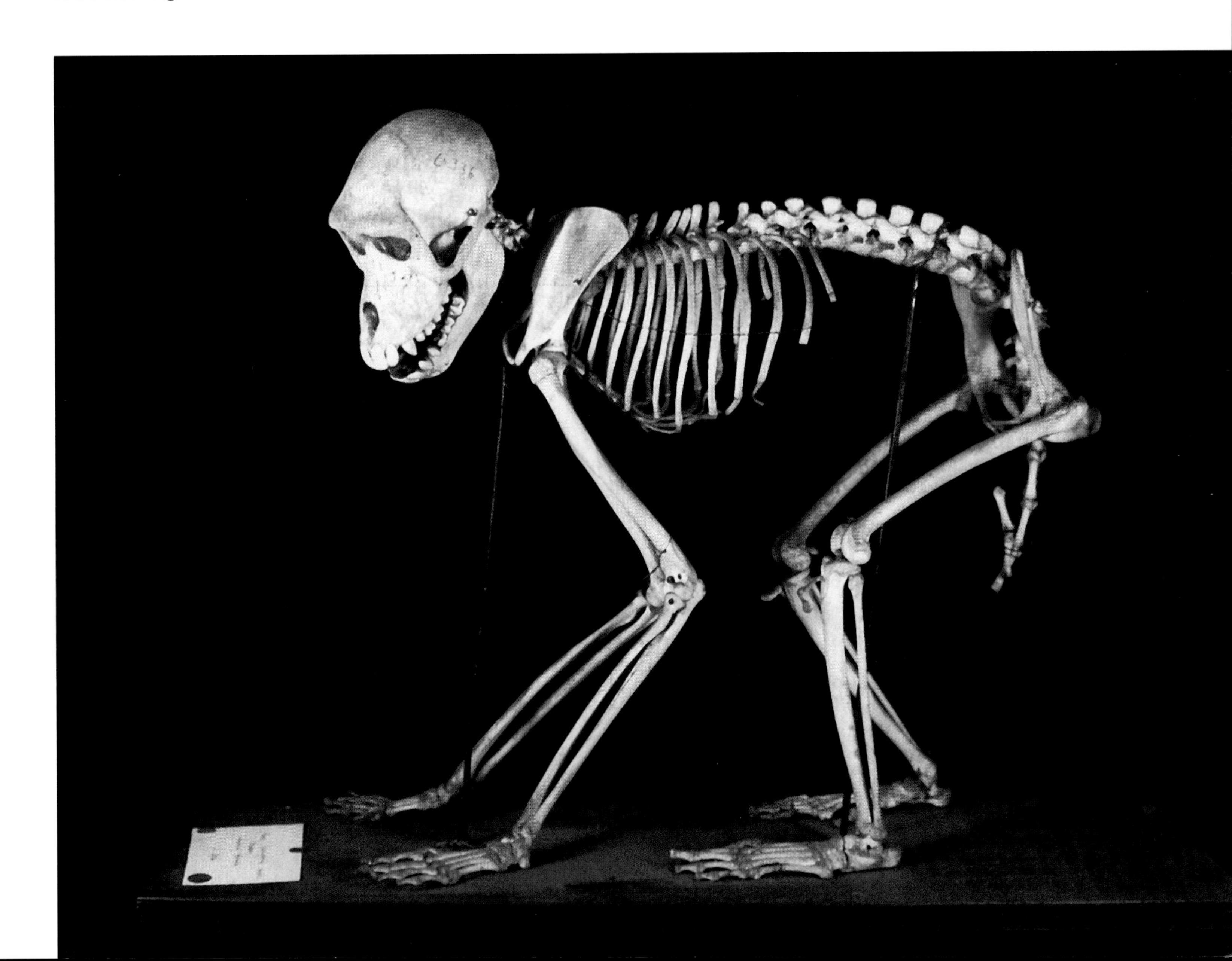

Overnight, the context for understanding "man's place in nature" changed—sending out shock waves that still reverberate. The message that we share an evolutionary history with the rest of the living world has yet to be fully assimilated into our ongoing collective attempt to understand exactly who we are. The late-nineteenth century rush to reinterpret fossils in the light of evolution naturally included the relatively scant human prehistoric remains that had already come to light. But there was a lingering reluctance on the part of many to concede that there really had been an evolutionary history to our own species, already dubbed by Linnaeus *Homo sapiens*, "man the wise." Thus so-called "human exceptionalism": many people continue to find that evolution of worms, trilobites and maybe even monkeys is OK—as long as humans are left out of the picture.

A skull discovered in a cave on the Rock of Gibraltar had generally been interpreted as pathological. [Plate 54] After Darwin's *On the Origin of Species* had appeared, and because similar skulls were turning up here and there around Europe, some biologists opined that the Gibraltar skull and those like it were actually fossil remains of our own evolutionary predecessors. But as late as 1872, the renowned German biologist Rudolf Virchow reverted to the older interpretation: these fossils seemed to him to be the remains of diseased modern individuals. The road to full acceptance of our own evolutionary interconnectedness with the rest of life was a bumpy one indeed.

But cling as some biologists would to older notions of human pedigree, it was inevitable that the general reinterpretation of the fossil record in the light of evolutionary theory would embrace the human lineage. In 1859, Darwin had merely (and modestly) suggested that "descent with modification" would eventually "shed light" on the question of human origins. But, by 1871, he brought the matter to stark attention in his *The Descent of Man*. In that spirit, a Dutch physician, Eugene Dubois, determined to find fossils of early man—the "missing link" between apes and modern humans. Darwin logically suggested that such connections were to be found in Africa, where most of the great apes are still to be found. But Dubois lacked the personal resources to travel to Africa. Instead, as the next best thing, he joined the Dutch East India Company, setting out in 1887 for Sumatra, the home of orangutans, the Asian great ape. [Plate 55]

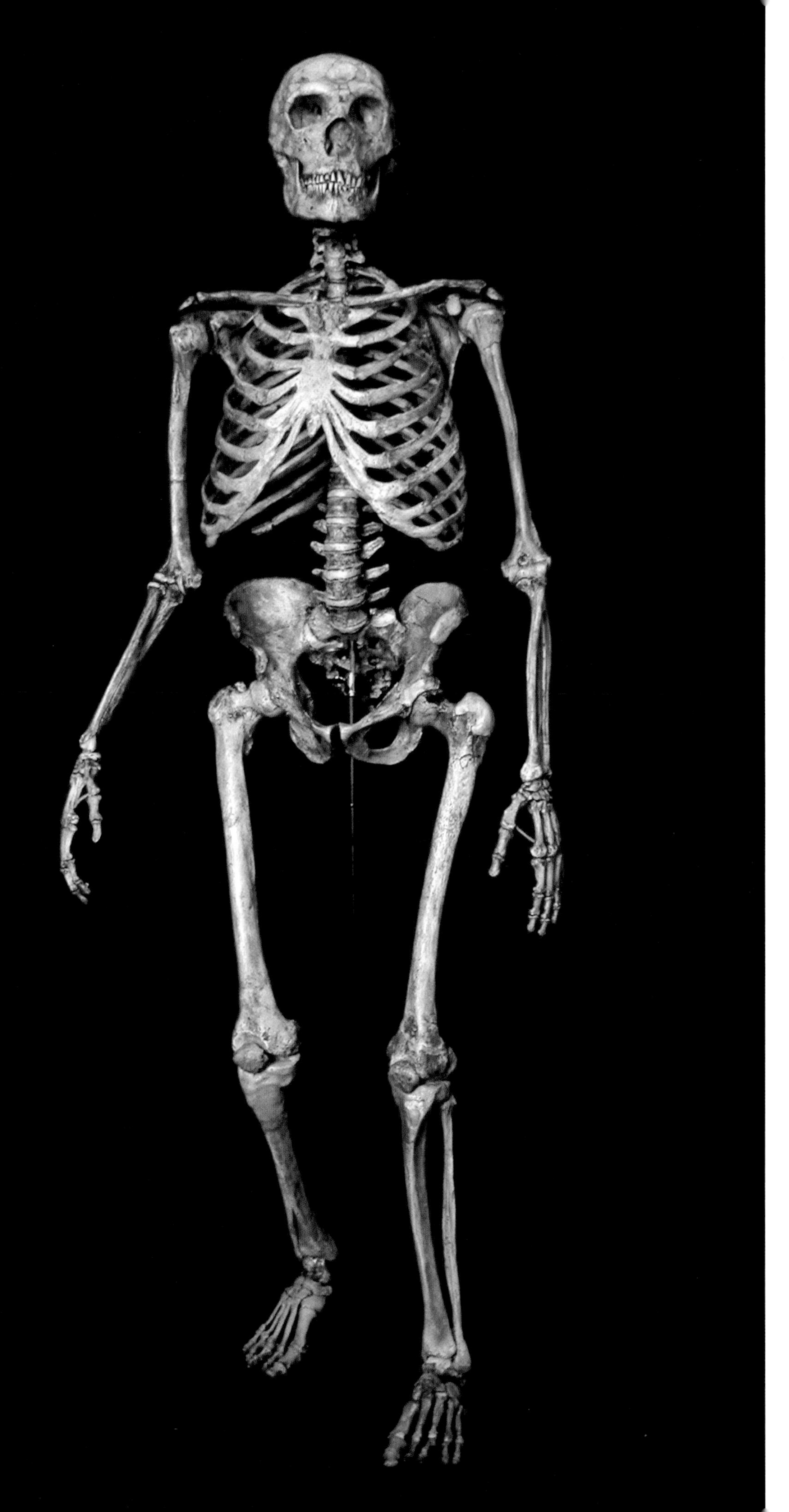

PLATE 54

HOMO NEANDERTHALENSIS,
RECONSTRUCTED SKELETON.

The remains of over 400 Neanderthals have been unearthed. This reconstruction was built by modeling the bones of eight of the most complete of these fossil Neanderthal skeletons.

Credits: Reconstruction by G.J. Sawyer, American Museum of Natural History, and Blaine Maley, Washington University; Photograph by Kenneth Mobray, CyraCom International.

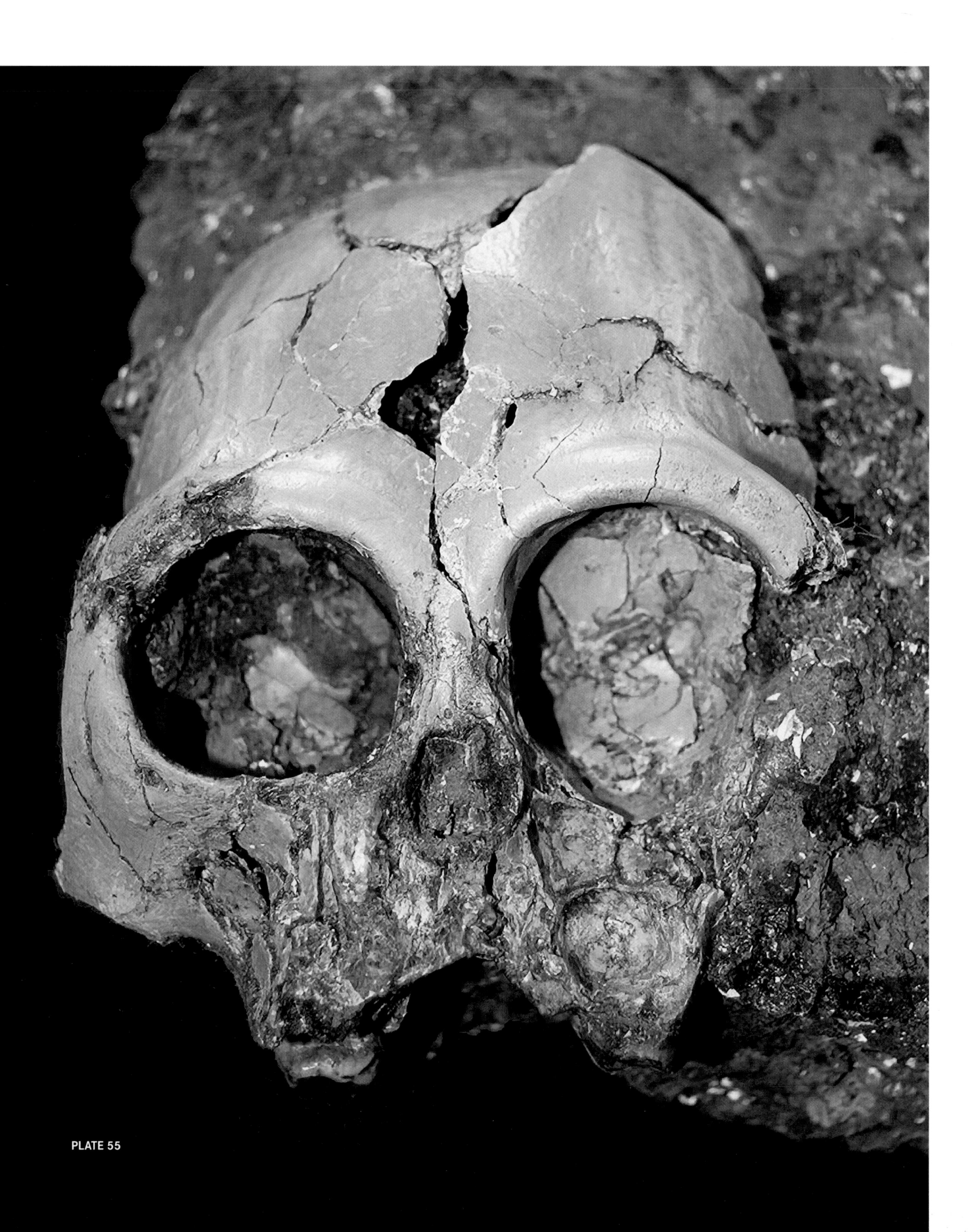

PLATE 55

PLATE 56

Dubois and his team of helpers struck paydirt remarkably quickly. By 1891, in central Java, bones and teeth of what he called *Pithecanthropus erectus* were beginning to turn up. [Plate 56] The name means "erect ape man" and announces Dubois's conviction that he indeed found the bones of a species intermediate between ape and man. Though these creatures walked erect (abundantly clear from the shape of the femur, or thigh bone, which is very similar to our own), their rather thick skulls had massive brow ridges and, most importantly, housed a brain only 70% the size of our own.

PLATE 56

HOMO ERECTUS*. FEMUR AND SKULL CAP, ORIGINALLY KNOWN AS *PITHECANTHROPUS ERECTUS

Middle Pleistocene (*ca.* 1 million years). Trinil, Java.

These are among the fossil unearthed by the Dubois team in the 1890s.

PLATE 55

***LUFENGPITHECUS LUFENGENSIS,* SKULL *IN SITU*.**

Upper Miocene (*ca.* 6 million years ago). Shuitangba, Zhaotong, Yunnan Province, People's Republic of China.

This beautiful skull is considered to represent a species within a lineage of large apes not especially closely related to any of the living great ape species of today's world.

Credit: Ji Xueping, Yunnan Institute of Cultural Relics and Archaeology.

Dubois put the infant science of human paleontology on a firm footing. Though he himself was later to waver in the face of severely adverse criticism (Dubois later dismissed his fossils asmerely those of an extinct species of gibbon), the world has long since accepted his fossils as the authentic remains of a farflung extinct species of fossil human, now renamed *Homo erectus*. The species lived throughout the Old World, and its bones and tools are known from southern Africa all the way over to China and those Javanese deposits first explored by Dubois and his crew.

Through the intervening years, many names have been added to the pantheon of successful hunters of human fossil remains. In the 1920s Raymond Dart brought Africa into the fold of productive sites, finally confirming Darwin's prediction with his discoveries of species still older and more primitive than Dubois's *Pithecanthropus erectus*. Dart, too, faced opposition to his interpretation. Today we accept these early African fossils that Dart had christened *Australopithecus africanus* for precisely what Dart had originally taken them to be: exceedingly early and primitive members of the human lineage. [Plate 57] They are so primitive, in fact, that they do indeed forcefully remind one of apes. The farther back in time we extend our search, the more ape-like our progenitors seem to have been. Just as we would expect, if humans really had diverged from the ape lineage in the remote past.

PLATE 57

***AUSTRALOPITHECUS AFRICANUS,* SKELETON UNDERGOING EXCAVATION.**

Dubbed "Little Foot," this hominin walked bipedally, but also spent a significant amount of time climbing in trees.

Credit: Courtesy Ron Clarke, University of the Witwatersrand.

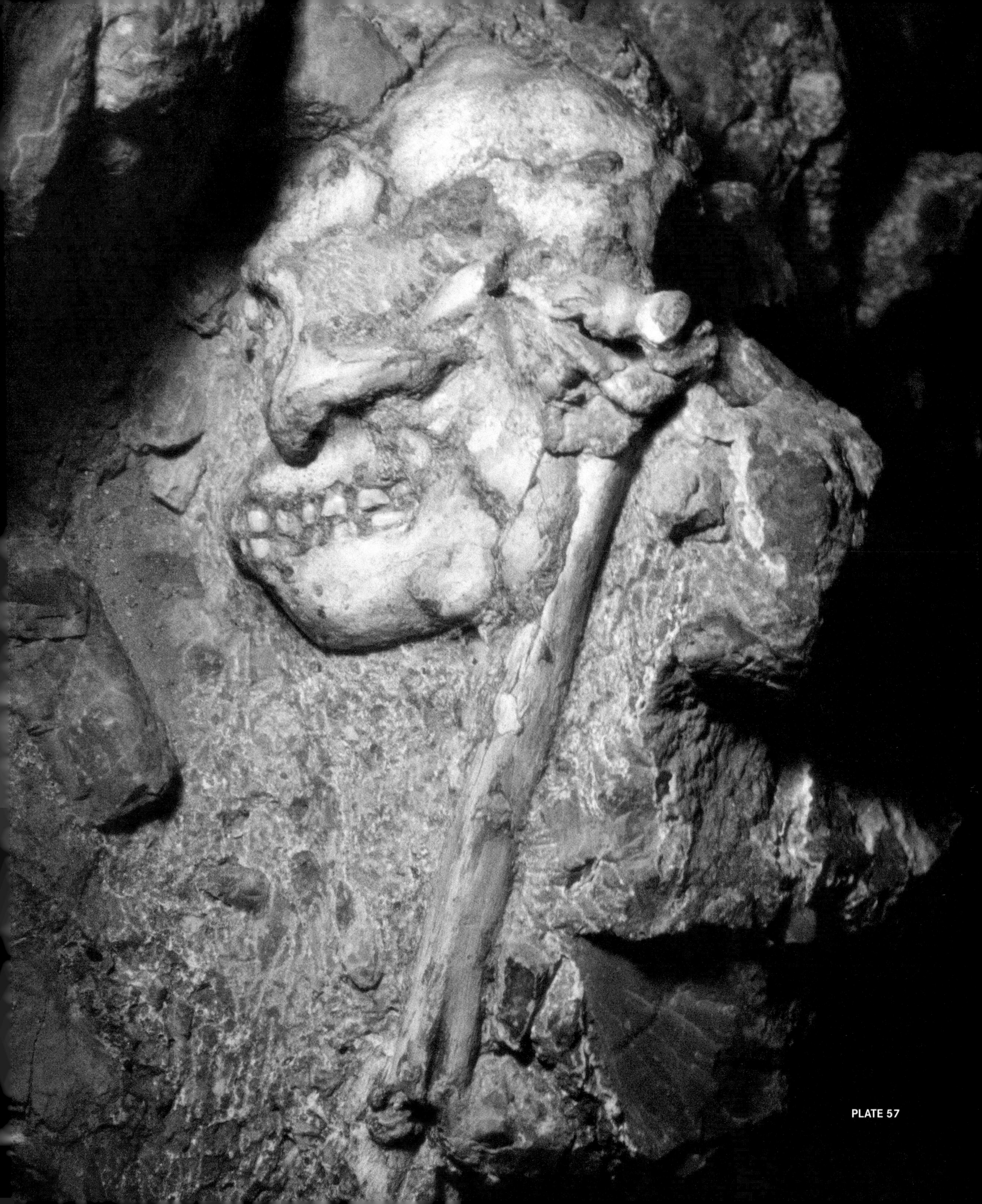

PLATE 57

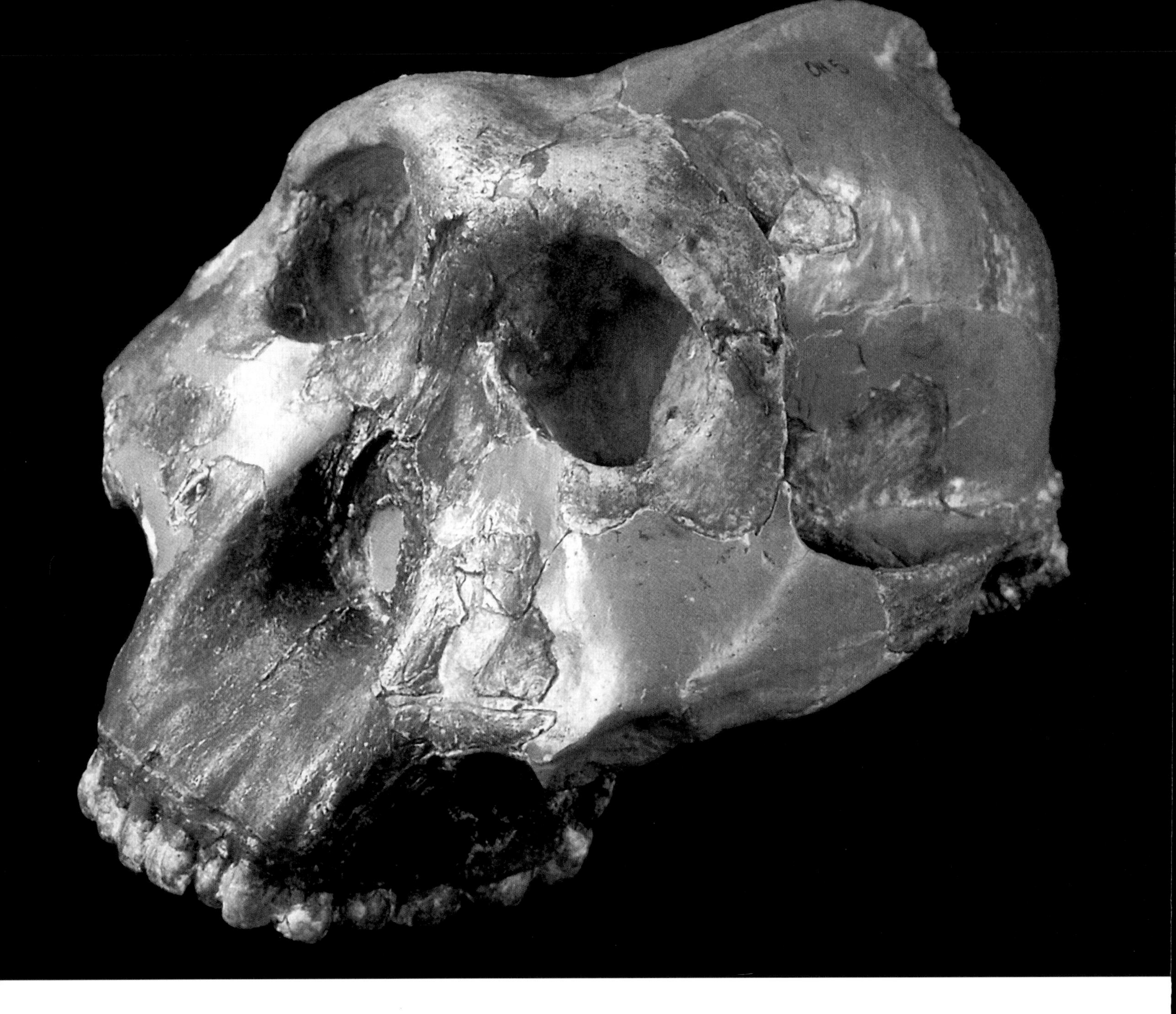

PLATE 58

***PARANTHROPUS BOISEI,* ROBUST AUSTRALOPITHECINE ORIGINALLY KNOWN AS *ZINJANTHROPUS BOISEI*.**

This is the skull that marked the Leakey's first success, after many years spent prospecting for early human remains at Olduvai Gorge and other East African locales.

Famous, too, are many of the later field leaders of successful exploratory ventures in human paleontology. Most famous of all are the several Leakeys. Louis S. B. Leakey and his equally renowned archeologist wife, Mary Leakey, possessed a dogged determination that kept them persisting through many disappointing years at Tanzania's Olduvai Gorge. Finally, in 1959, patience was rewarded when Mary found the skull of *Zinjanthropus boisei*, today reclassified as *Paranthropus boisei*. [Plate 58] Mary Leakey's discoveries of the fabulous trackways of primitive human footprints at Laetoli, in Tanzania, confirm that the most ancient of our known progenitors were upright and walking as long ago as 3.5 million years.

More recently, the second generation of Leakeys has made profound contributions to human paleontology. Richard Leakey, flying south into Kenya from an exploration site in Ethiopia, saw from his helicopter promising beds exposed around Lake Turkana in northern Kenya. His finds of specimens of *Australopithecus*, and especially of early members of the genus *Homo*, including *Homo habilis* and the first known specimens of African *Homo erectus* (known as *Homo ergaster*), have greatly enriched our picture of human evolution. [Plate 59]

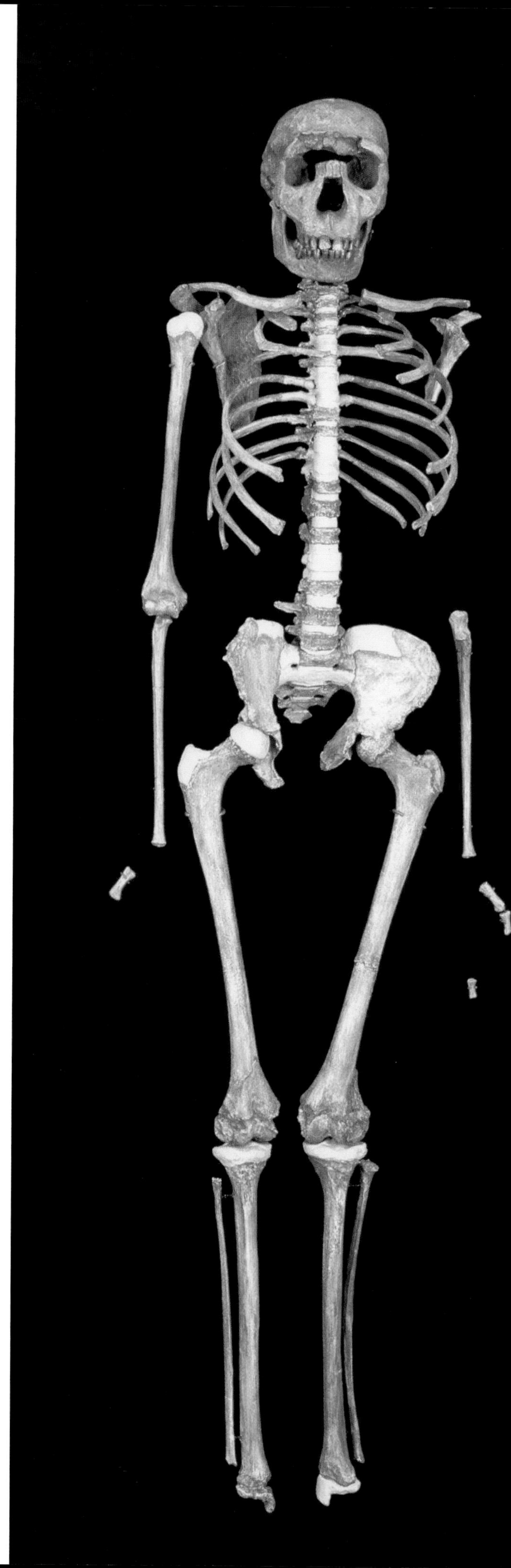

PLATE 59

HOMO ERGASTER,
TURKANA BOY SKELETON.

Below the neck the body proportions of this 8-year old hominin were essentially like our own. His brain capacity at 880 cc., however, was considerably smaller than modern humans who average 1350 cc.

Credit: Courtesy Ian Tattersall, American Museum of Natural History.

It is important to acknowledge the work of these intrepid explorers of human history. Their fame flows directly from the importance that humanity at large attaches to understanding our history. As the only (so far as we know) species that is actively conscious of its own existence, we are privileged to wonder at the very nature of being alive. The very human squabble between Richard Leakey and Donald Johanson, famed discoverer of "Lucy" and other Ethiopian specimens, one of the very oldest known species in the human line of ancestry, over details of the course of human evolution, reflected the importance attached to every nuance of interpretation, each subtle disagreement between two titans of human fossil collecting. As humans, it is only natural that we attach such special importance to the fossils that reveal the nature of our own evolutionary history.

The disagreement that seems to be such a constant fixture in human paleontology exaggerates, in truth, the actual difficulties in discerning the basic outline of our evolutionary history. Thanks to these and other indefatigable collectors, the human fossil record has become sufficiently dense to provide a picture that is at least as clear as the evolutionary histories of the vast majority of other lineages known from the fossil record. We will continue to debate details and revise the precise topology of the family tree, but, in the main, the overall picture is unlikely to be radically altered.

AN OUTLINE OF HUMAN EVOLUTION

Darwin certainly seems to have been right: fossils bear out his contention that our lineage really did arise in Africa. Discovered in the Djurab Desert of Chad of Central Africa, *Sahelanthropus tchadensis* lived around 7 million years ago, making it the oldest known human ancestor after the split of the human lineage from that of chimpanzees. (See Plate 70.) Remains of *Orrorin tugenensis* at 6 million years were found in Kenya. [Plate 60] Both species evidence anatomical features suggesting that they were at least partially bipedal. And some have claimed that both had greater affinities to *Homo sapiens* than the lineage of early humans represented by the younger species of *Paranthropus* found in southern Africa and by the Leakeys at Olduvai.

Best known of early ancestral humans, Lucy and her cohorts, come from east Africa, especially from Ethiopia and Laetoli, Tanzania. In many respects, this species—*Australopithecus afarensis*—shows little real differentiation from great apes. Lucy herself is an especially important find. Named for the Beatles' song popular at the time of her discovery, this fossil is the incomplete remains of an entire (and apparently female) skeleton—something of a rarity in the human fossil record, where teeth and disconnected portions of skeletons are the usual rule. [Plate 61] The partial skeleton tells us some fundamentals of size of *Australopithecus afarensis.*

PLATE 60

***ORRORIN TUGENENSIS,* FEMURS (TOP LEFT AND RIGHT) OF AN EARLY HOMININ.**

Late Miocene (*ca.* 6 million years). Lukeino Formation, Tugen Hills, Kenya.

The morphology of this femur is more human-like than those of the australopithecines or African apes.

Credit: Photograph by Martin Pickford, Collège de France; courtesy of Community Museums of Kenya.

PLATE 60

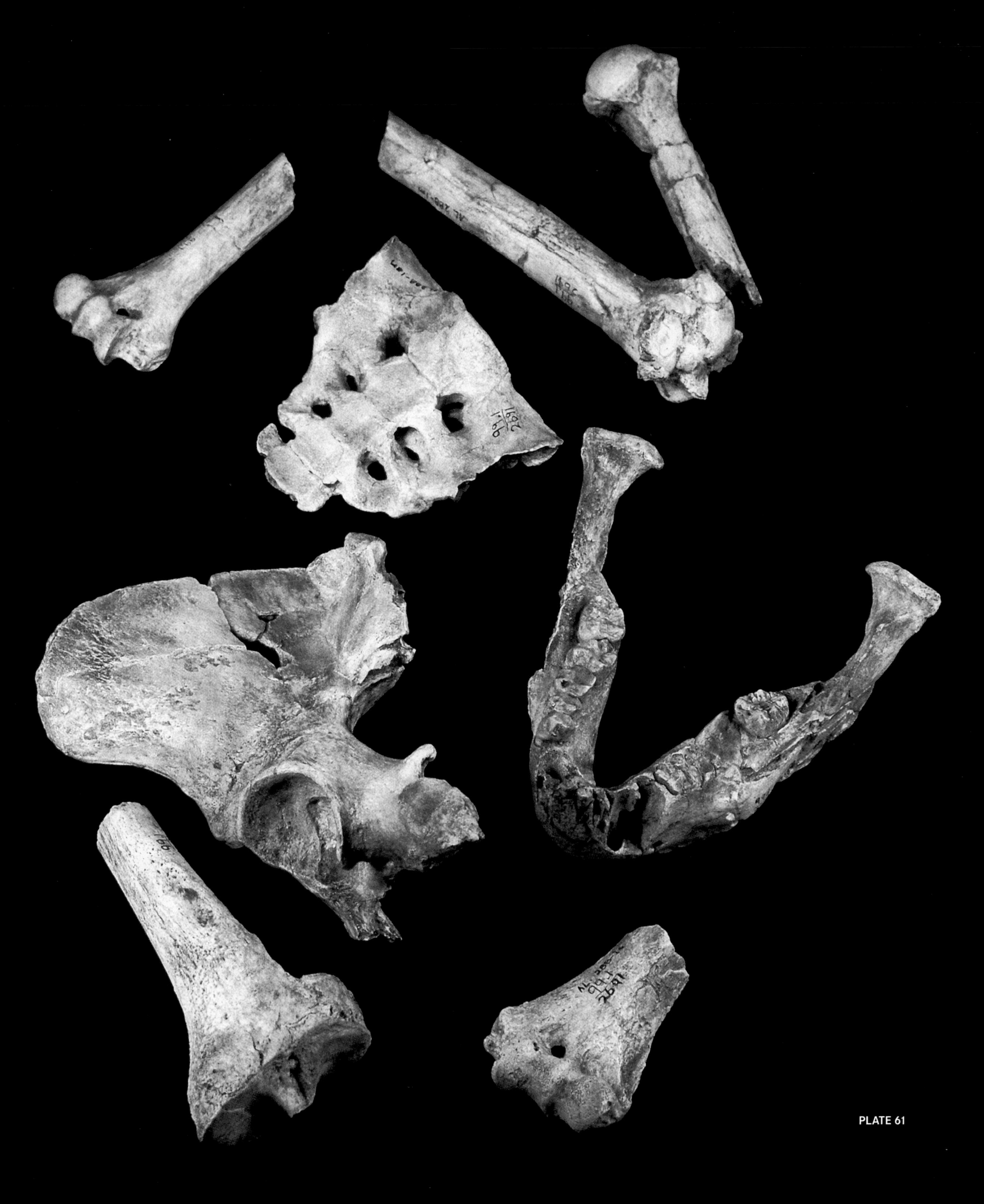

PLATE 61

Standing at about four feet tall, Lucy was about the size of a modern chimpanzee. More to the point, the brain in this species, estimated to have averaged 450 cubic centimeters (cc.) in volume, hardly exceeds the cranial capacity of modern day chimps. Many other features are distinctly ape-like, but show some slight modification in "our" direction: the jutting face houses canine teeth that are large by our standards, but reduced from the ape condition. [Plate 62] Like later members of our lineage, this species' chewing teeth (molars) were already a bit enlarged. But the signal difference between apes and these early members of the human lineage is their bipedalism: as we have seen, Mary Leakey's footprints at Laetoli confirm without a shadow of doubt that *Australopithecus afarensis*, however otherwise primitive and ape-like, however much it may still have lived in trees, was fully capable of upright posture and sustained two-legged locomotion. In recent years, traces of Lucy's own ancestry have started to yield to the probing efforts of paleontologists.

And, as the intense search for our own ancestors continues, new discoveries in Africa have revealed traces of other species that were living contemporaneously with *A. afarensis.* This was confirmed by a 1999 discovery in Kenya of a 3.5 million year old hominin dubbed *Kenyanthropus platyops* by a team led by Meave Leakey, along with her daughter Louise today the most active members of the fabled Leakey clan. [Plate 63] Though paleoanthropologists are famous for their disputes on the evolutionary relationships and proper classification of newly uncovered fossil hominins, evolutionary biologists have long been pointing out that much of evolution proceeds in isolated populations leading to new species. It should come as no surprise, in the five-seven million year history of our lineage, that there were often more than one species alive at any one time. It is very much as if we are biased by the fact that only one species—our own species, Homo sapiens—is alive in the modern world. The fossils show clearly that there were usually at least two, or more, species in the lineage—an absolutely to-be-expected condition of any actively evolving lineage.

PLATE 61

***AUSTRALOPITHECUS AFARENSIS.* SOME OF THE SKELETAL BONES OF LUCY.**

Upper Pliocene (*ca.* 3.2 million years). Hadar, Ethiopia.

Found by Donald Johason, "Lucy" is one of the most famous fossil finds in modern paleontology.

PLATE 62 **(See page 102.)**

***AUSTRALOPITHECUS AFARENSIS,* UPPER AND LOWER JAWS.**

The jaws of *A. afarensis* house teeth distinctly more human than ape-like.

PLATE 63 **(See page 103.)**

***KENYANTHROPUS PLATYOPS,* A MEMBER OF THE HOMININI, A GROUP THAT INCLUDES MODERN AND EXTINCT HUMANS AFTER THEIR SPLIT FROM CHIMPANZEES.**

Pliocene (*ca.* 3.35 million years). Lake Turkana, Kenya.

Kenyanthropus platyops shows that at least two species (the other being *Australopithecus afarensis*) of hominins existed in eastern Africa at the same time more than 3 million years ago.

Credit: Photograph by Fred Spoor, Max Planck Institute of Evolutionary Anthropology; courtesy of National Museums of Kenya.

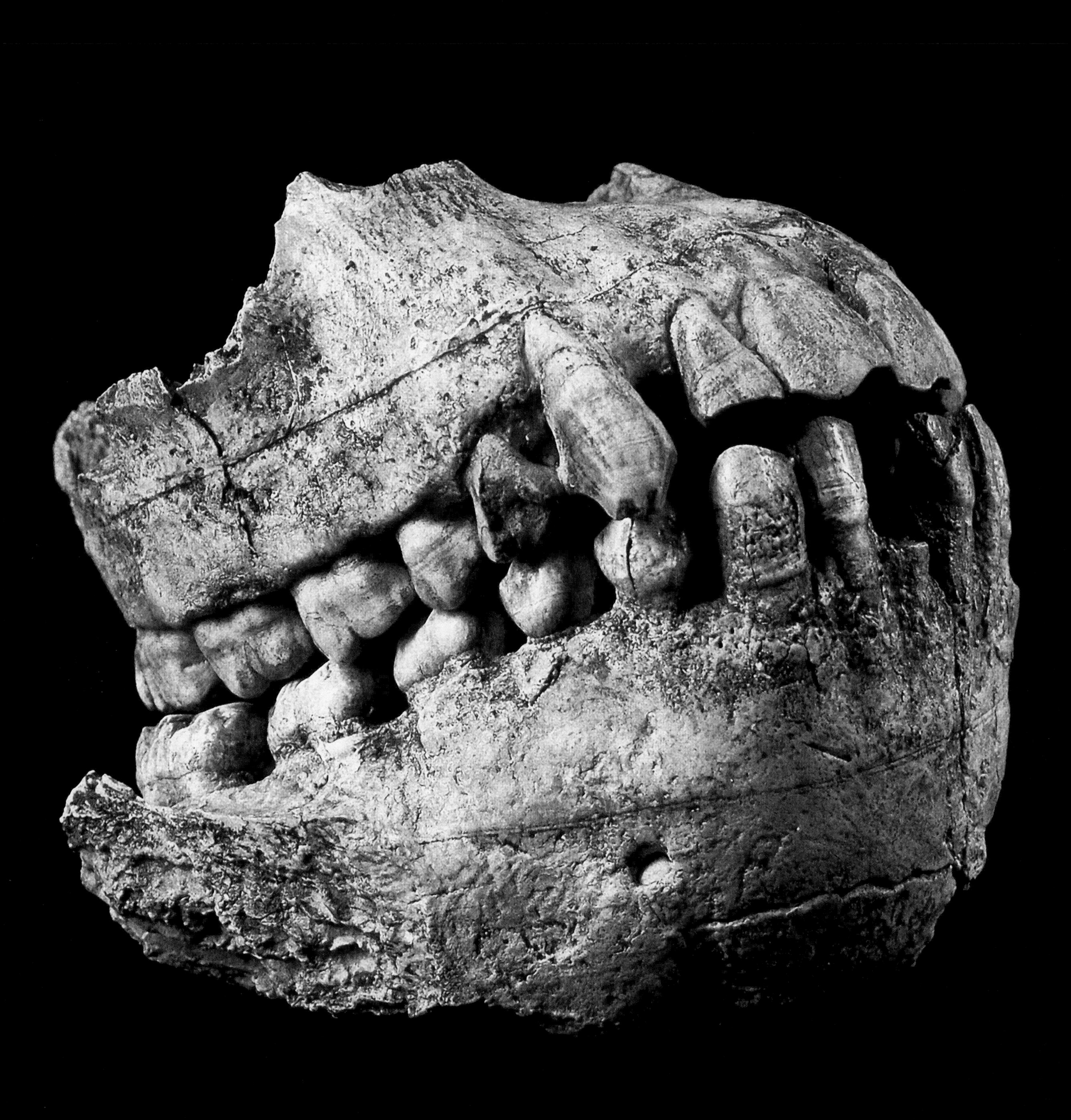

PLATE 62

PLATE 63

Lucy and company lived 3–4 million years ago. Next in line, 3–2.5 million years old, are fossils of Dart's species, *Australopithecus africanus*. Known primarily from limestone caves in present-day South Africa, these creatures were more lightly built and a bit less apish than *Australopithecus afarensis*. Yet their brains, on average, seem no larger than the 450 cc. brains of *afarensis*. Their slightness of build (they were "gracile" in the terminology of paleoanthropology) above all else stamps them as members of our lineage, advanced ever so slightly over their earlier African forebears.

Things changed abruptly in Africa about 2.5 million years ago. The woodlands and tropical forests rather quickly gave way to vast open stretches of grasslands and savannah, with isolated pockets of woodlands scattered around as islands dotting the plains. The result of a rather sudden drop in global temperature, this abrupt reorganization of habitat spelled the death knell for large numbers of species. Though there is some suggestion that *Australopithecus africanus* might have lingered on in southern Africa, it is right after these dramatic climatic events that the very recently discovered 2 million year old *Australopithecus sediba* skeleton enters the picture. For the first time, just after 2.5 million years ago, we find not one, but two species (some paleontologists say three) of proto-humanity living at the same time.

The first of these two is an offshoot—at first glance, superficially back in the direction of apes. Some species had become secondarily thick-boned—so-called "robust" *australopithecines*, such as the South African species *Paranthropus aethiopicus* and *Paranthropus robustus* and Mary Leakey's *Paranthropus boisei*. This complex of robust *australopithecine* species lasted about 1 million years, living alongside the second species: *Homo habilis,* the "handy man." So named for the very human-like hand bones discovered by Louis Leakey at Olduvai Gorge, it is here that we see a major step in our direction, anatomically speaking. The famous skull of *Homo habilis* discovered by Richard Leakey had a cranial capacity of about 750 cc.—still far less than our average of 1400 cc., but a major increase over the capacities of the earlier species nonetheless. [Plate 64] There is evidence, moreover, that this "handy man" was the first to fashion stone tools, which coincidentally begin to appear in the archeological record just when *habilis* bones show up, beginning at about 2.6 million years ago. True to form, this precedence has been challenged by what are claimed to be 3.4 million year evidence of tool use discovered at Dikika, Ethiopia and associated with Lucy and her kin. [Plate 65]

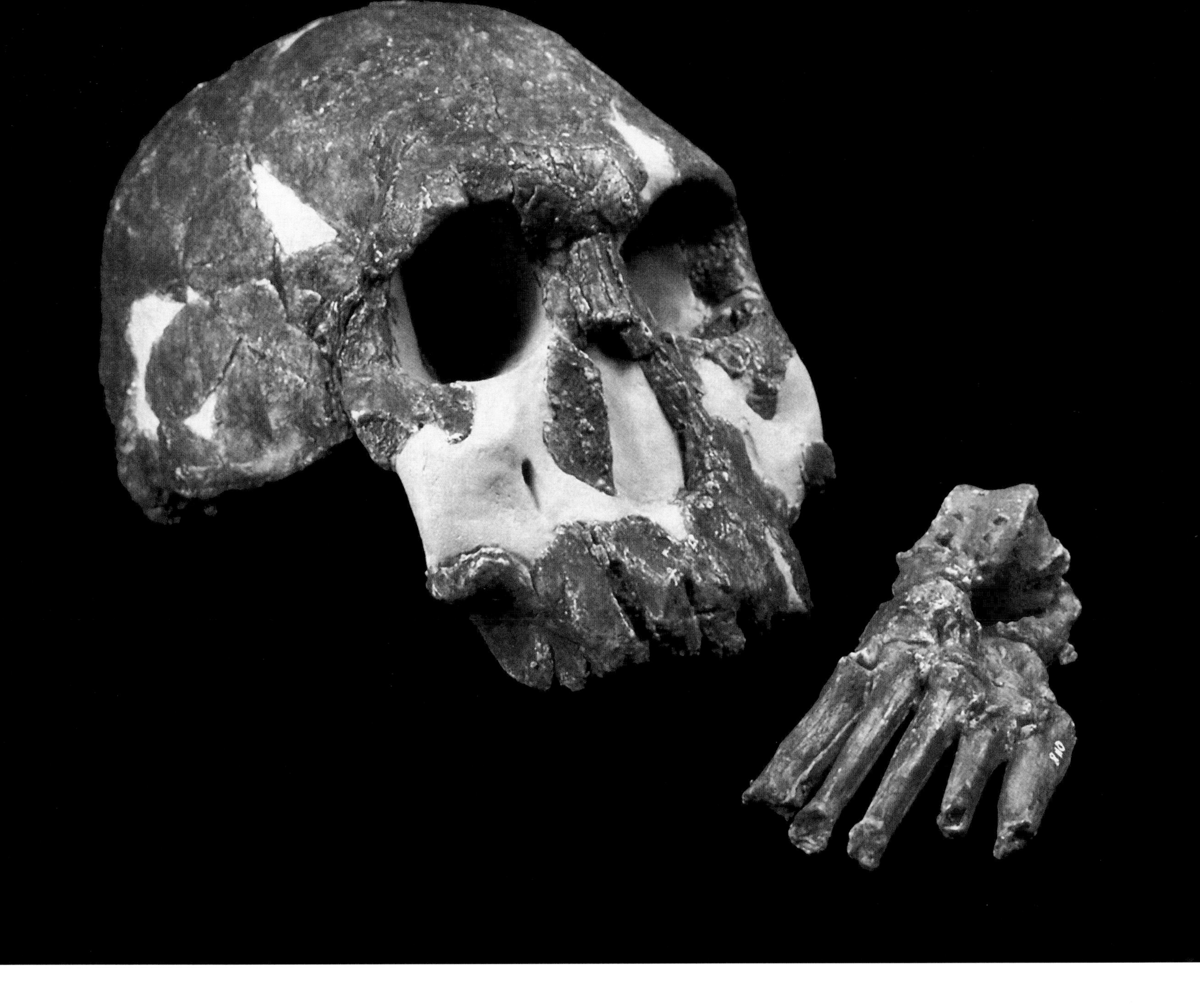

PLATE 64

***HOMO HABILIS,* LEFT: FAMOUS SKULL ER1470.**

Upper Pliocene (*ca.* 1.9 million years). Koobi Flora, Kenya.

RIGHT: FOOT BONES.

Upper Pliocene (*ca.* 1.8 million years). Olduvai Gorge, Tanzania.

The brainier skull, along with foot bones, of "handy man"—oldest species yet to be classified in our own genus: *Homo*.

PLATE 65

IMPALA-SIZED RIB AND BUFFALO-SIZE FEMUR FRAGMENTS WITH CUT MARKS.

Middle Pliocene (*ca.* 3.4 million years ago). Dikika, Lower Awash Valley, Ethiopia.

Cut marks are indisputable products of human tool-use behavior.

Credit: Copyright© Dikika Research Project, courtesy Shannon McPherron, Max Planck Institute for Evolutionary Anthropology.

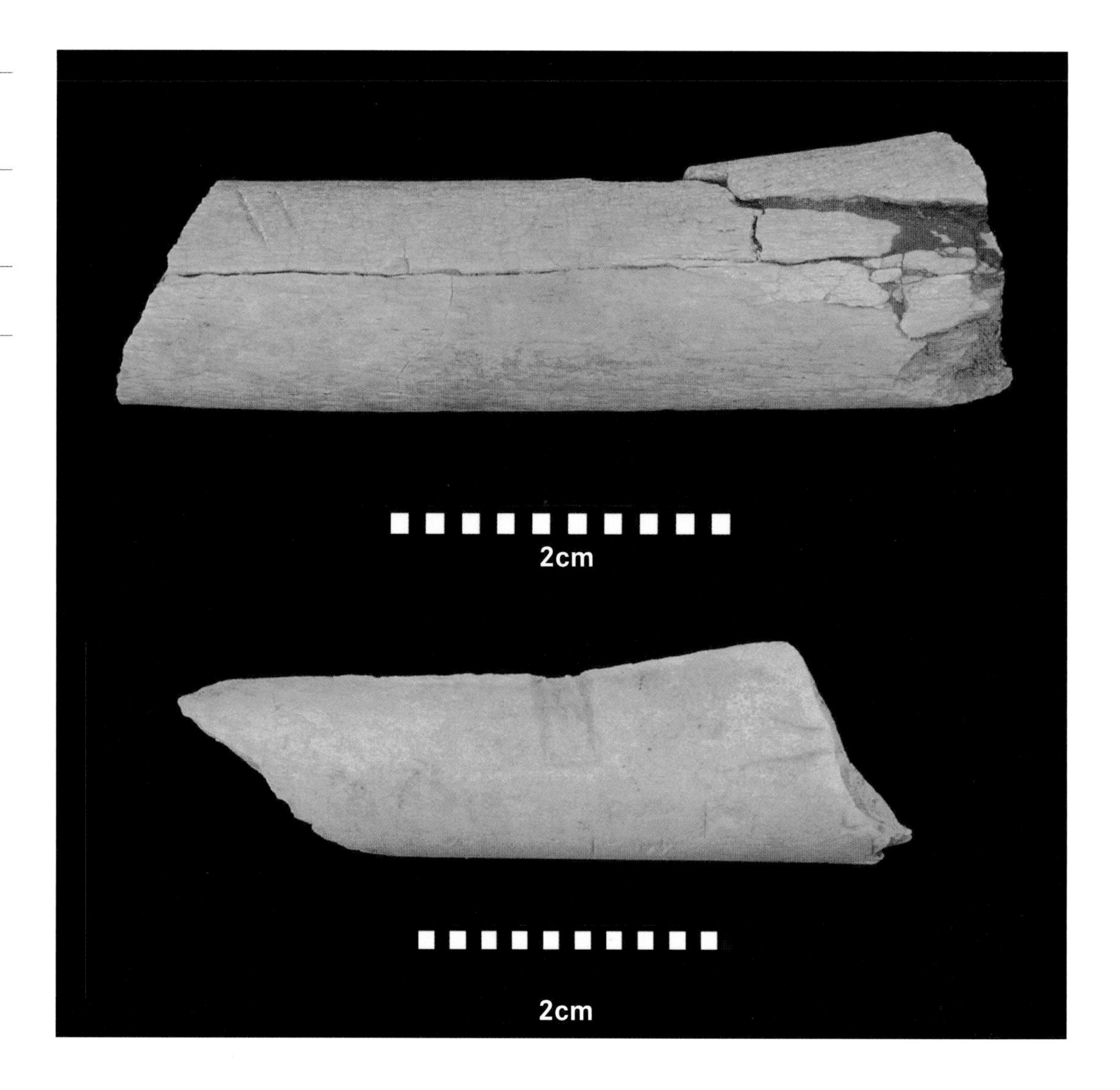

PLATE 66

***HOMO NEANDERTHALSIS*, FEMALE TOE BONE.**

Middle Pleistocene (*ca.* 130,000 years). Denisova Cave, Siberia.

Amazingly, this toe bone, interpreted to have belonged to a Neanderthal female, has yielded a nearly complete genome of someone living 130,000 years ago. Modern humans split off from Neanderthals ca. 600,000 years ago; genomic studies such as this one have shown that 1% to 3% of the DNA of modern Europeans and Asians comes from Neanderthals, most likely from gene exchange after the lineages had already diverged.

Credit: Bence Viola, Max Planck Institute for Evolutionary Anthropology.

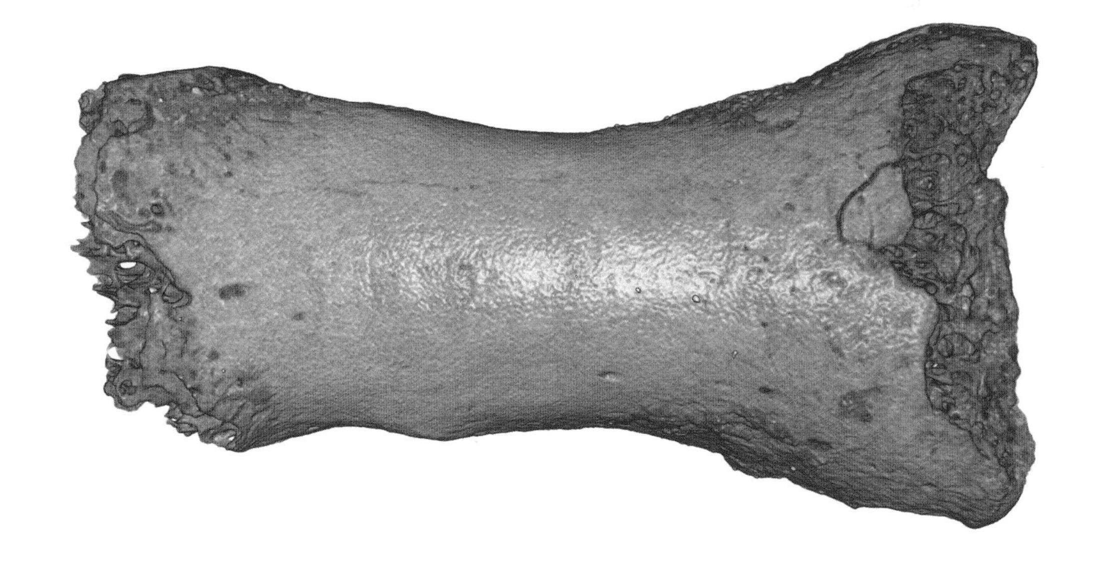

So there was speciation in our history, with one line, *Homo,* leading to ourselves, the other, robust species of *Australopithecus*, persisting as vegetarian specialists for about 1 million years before yielding to extinction. Next in line was *Australopithecus sediba* transitional between the gracile *australopithecines* and Dubois's Asian species *Homo erectus* and its close relatives, such as the earlier *Homo ergaster* in Africa. Thanks to the 1984 discovery of a 1.6 million year old skeleton along the shores of Kenya's Lake Turkana African, as well as the large femur Dubois had collected in Java in the 1890s, we know that *ergaster/erectus* was nearly as tall as we are. [See Plate 51.] The earliest known specimens are nearly 2 million years old, and slightly predate the first glacial pulse that began the Pleistocene Epoch ("Ice Age"). The youngest specimens so far discovered, from a cave near Beijing, China, are only 300,000 years old.

Homo erectus, then, was a very successful and eventually far-flung species that lived for well over a million years. Brain size averages around 1000 cc. in this species—and, remarkably seems to have changed but little during the entire 1.5 million year history of this species. Some anthropologists believe that there is a slight increase in brain size from oldest to youngest specimens, but others point to the rather high value of 1067 cc., known from an African skull that is one of the very oldest specimens yet found of the *ergaster-erectus*. With its pronounced brow ridges below a flattened, sloping forehead, erectus would have looked quite primitive to our eyes. Yet this species was otherwise quite human in appearance, and apparently in behavior as well: they had stone tools, hunted and presumably fashioned clothing.

Until recently, the fossil hominin finds have been few and far between in Europe. Especially striking are the five *ergaster-erectus* skulls excavated in the Republic of Georgia. They are dated at 1.8 million years, *i.e.* are roughly as old as the oldest *ergasters* known from Africa. My colleague at the American Museum of Natural History, anthropologist Ian Tattersall, has long believed there was at least one speciation event towards the end of *ergaster-erectus* history: *erectus* may have survived longer in Asia, replaced by a distinct, descendant species in Europe. The new Georgian finds show that even to have likely begun earlier than previously imagined. The rather sparse fossil record between 400,000 and 100,000 years ago in Europe shows a fair measure of variation. [Plate 51, see page 88.] In any case, by 100,000 years ago the Neanderthals, *Homo neanderthalensis* were firmly ensconced there. Some paleoanthropologists still prefer to classify Neanderthals as a subspecies of our own species, *Homo sapiens neanderthalensis,* but there can be little doubt that the differences between Neanderthals and modern humans are greater than the differences generally encountered between two closely related species. Though Neanderthal brains were, if anything, even slightly larger than our own (1500–1600 cc. compared to our average of 1400 cc.), they were organized somewhat differently: there is truth to the old picture of beetle-browed, sloping foreheaded people, though Neanderthals were by no means the stooped brutes that they had long been made out to be. The high foreheads of our own species, *Homo sapiens*, marks the presence of an expanded mass of the brain's cerebrum: we are literally far more cerebral than our Neanderthal cousins were. [Plate 66]

Once again the scientific infighting has flared over how many species there were outside of Africa starting just under 2 million years ago. A recent report claims that basically one variable species accounts for all the rich anatomical diversity that is found among the hominins of Eurasia in the Plio-Pleistocene. Members of the press are solemnly declaring that we must once again go back to the drawing boards and completely rethink our picture of our own evolutionary story. And once again, some paleoanthropologists are declaring their boredom of the eternal wrangle between the "lumpers" and the "splitters," i.e. those who tend to include wider ranges of variation in a single species, vs. those who see variation packaged in discrete species, especially when separated in space and in time.

Nonsense. This is not a dispute about classification, but rather the recognition of that packaging of morphological (hence genetic) diversity—the better to understand how the evolutionary process works—and specifically how it worked to produce the array of ancestral hominins—and ultimately ourselves. As climate fluctuated and huge ice sheets waxed and waned during the last million and a half years, it must be the case that separate lineages were evolving in what is now Europe and what is now Asia. [Plates 67, 68]

PLATE 67

***HOMO GEORGICUS*, ADULT HOMININ SKULL DOCUMENTING THE EARLY PRESENCE OF *HOMO* OUTSIDE AFRICA.**

Lower Pleistocene (*ca.* 1.77 million years). Dmanisi, Georgia.

The Dmanisi skulls prove that *ergaster/erectus* hominins reached Europe sooner than previously thought. Debate rages, however, on their significance:
Do they represent a speciation event that eventually initiated the Neanderthal lineage? Or do they confirm the traditional gradual view of human evolution passing through progressive stages more or less simultaneously throughout the Old World. My money is on the former possibility.

Credit: David Lordkipanidze, Georgian National Museum.

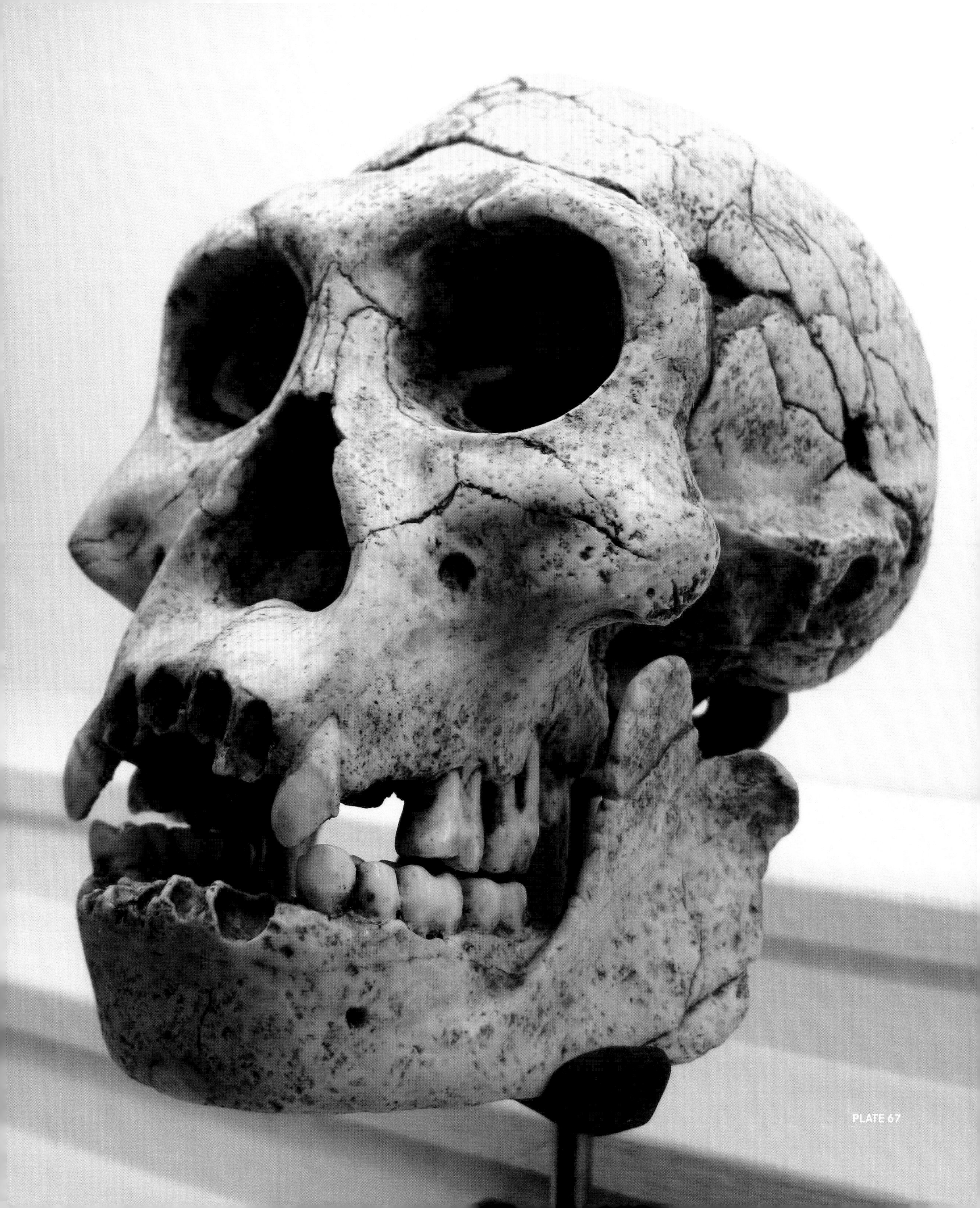

PLATE 67

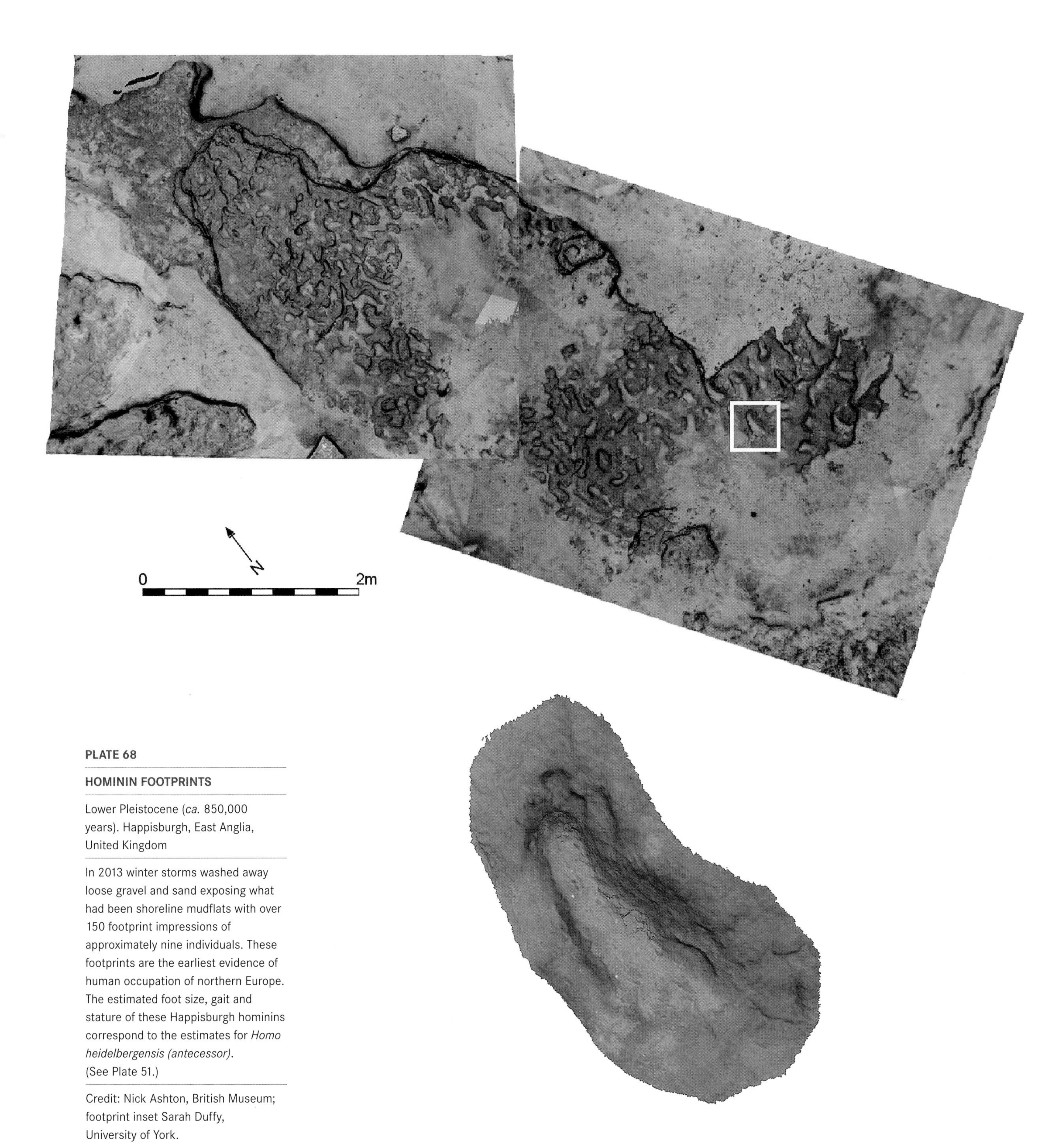

PLATE 68

HOMININ FOOTPRINTS

Lower Pleistocene (*ca.* 850,000 years). Happisburgh, East Anglia, United Kingdom

In 2013 winter storms washed away loose gravel and sand exposing what had been shoreline mudflats with over 150 footprint impressions of approximately nine individuals. These footprints are the earliest evidence of human occupation of northern Europe. The estimated foot size, gait and stature of these Happisburgh hominins correspond to the estimates for *Homo heidelbergensis (antecessor)*. (See Plate 51.)

Credit: Nick Ashton, British Museum; footprint inset Sarah Duffy, University of York.

The most graphic case in point: the case of *Homo floresiensis*. Discovered in cave deposits on the Indonesian island of Flores in 2003, these homins stood a mere three feet tall. Their diminutive crania house a brain that was around 380 ml.—or about the same size as the brain of a chimpanzee or *australopithecine*. They survived on Flores roughly from 100,000 years ago, disappearing a surprisingly scant 13,000 years ago. [Plate 69]

Naturally enough, debate has flared over exactly who these people were and how they came to be there. The *ergaster*-like quality of the skull is undeniable—yet some scientists have speculated that these were the remains of modern people (*Homo sapiens*) suffering from a condition known as microcephaly. Others have suggested various other genetic defects. But the simplest, and most compelling case, supported by an emerging majority of paleoanthropologists, is that Floresian man is a perfect example of island dwarfism. Indeed, there were two species of dwarf elephants on Flores at different time during the existence there of *Homo floresiensis*. Evolutionary dwarfism on islands is a common, if fascinating phenomenon. It seems to reflect diminished food resources, and perhaps other factors. So the betting is that here we have another example of genuine speciation off the *ergaster-erectus* lineage.

Meanwhile, our own species, *Homo sapiens*, with our high foreheads, well-developed chins and light ("gracile") skeletons had also arisen, around 200,000 years ago. Biochemical studies have predicted that the common ancestor of all of us would prove to have been African—a woman appropriately dubbed "Eve" living perhaps 140,000–200,000 years ago. (A woman because the study was based on mitochondrial DNA—and mitochondria, the power plants of all animal cells, are inherited strictly from one's mother.) Prediction confirmed: the oldest known fossils of our own species are approximately 190,000 years old, and are indeed African. [Plate 70] *Homo sapiens* reached Europe 41,000–44,000 years ago, our arrival coinciding with the demise of Neanderthals, and I am partial to the suggestion that we played a direct role in the extirpation of this separate species. Around 30,000 years ago, cave art of such exceptional esthetic value demonstrates that the full human creative potential had already been realized.

PLATE 69 **(See page 112.)**

***HOMO FLORESIENSIS*, "FLORESIAN MAN."**

Upper Pliestocene (*ca.* 18,000 years ago). Liang Bua, Flores, Indonesia.

This diminutive species, with its unmistakably primitive features, seems like a perfect example of "island dwarfism," the phenomenon affecting not only this primitive, pre-*Homo sapiens* hominin species living on Flores, but also at least two species of elephants living there as well.

Credit: Peter Brown, University of New England.

PLATE 70 **(See page 113.)**

***HOMO SAPIENS IDALTU*, AN ARCHAIC *HOMO SAPIENS* DISCOVERED IN 1997.**

Upper Pleistocene (*ca.* 160,000 years). Herto Bouri, Middle Awash, Ethiopia.

Despite its archaic features, this subspecies is thought to represent the direct ancestor of modern humans.

Credit: Berhane Asfaw, Rift Valley Research Service.

PLATE 69

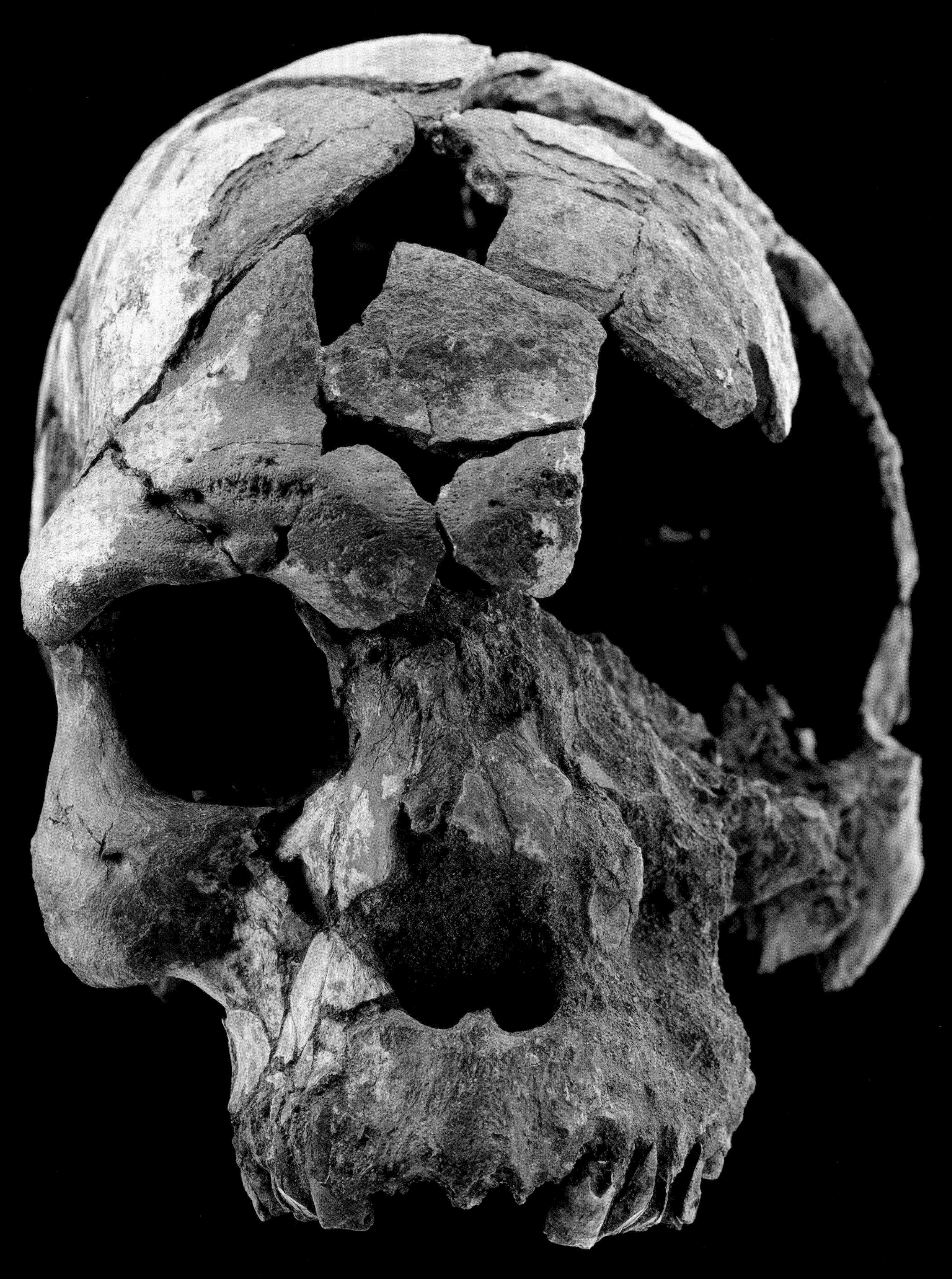

PLATE 70

PLATE 71

***SAHELANTHROPUS TCHADENSIS,* SKULL OF EARLIEST KNOWN HOMININ.**

Upper Miocene (*ca.* 7 million years). Toros-Menalla, Djurab Desert, Chad.

This early human ancestor lived very close to the time of the chimpanzee-human divergence.

Credit: Michel Brunet, Collège de France.

THE HUMAN PEDIGREE AND THE EVOLUTIONARY PROCESS

So much for the bare bones of human evolution. It is clear that the human line of descent is as much a product of the evolutionary process as is the history of all other organisms. But what does our own history have to tell us of the very nature of the evolutionary process? Does human evolution conform more closely to the Darwinian picture of slow, steady, gradual change, the progressive accumulation of unidirectional anatomical modification? Or does it look more like the picture of punctuated equilibria, with species appearing rather abruptly, and persisting for long periods with little or no change? The answer to these questions sheds light on the very nature of long-term accumulation of adaptive evolutionary change.

The human brain, as one might suspect, differs from ape brains in many ways. All one can reliably judge with fossils, however, is brain size: both absolute size (measured, say, in cubic centimeters) and relative size, *i.e.* brain size in relation to body size (as measured, for example, in pounds or kilograms). Anyone comparing contemporary humans and apes would conclude that, since their mutual divergence from a common ancestor 6–7 million years ago, brain size in humans has increased both absolutely and relatively far more than in the ape lineage.

This observation yields the important prediction that there should be a progressive trend towards brain size increase detectable in the fossil record of the human lineage. And that is precisely what we, in fact, see: the oldest member of our lineage, *Sahelanthropus tchadensis*, had brains no bigger than chimps (320 cc. to 380 cc.) and *Australopitecus afarensis* only slightly larger (380 cc. to 430 cc.). [Plate 71] Skipping over the robust australopithecine side branch, next comes *Homo habilis* (750 cc.), then *Homo erectus* (averaging 1000 cc.), then (skipping over the Neanderthal side branch) ourselves, at 1400 cc. Once again, prediction confirmed: there is a definite trend towards both absolute and relative increase in brain size revealed in the fossil record of human evolution. Ape brain size, meanwhile, seems to have stayed frozen below 500 cc. over the same interval of time.

The Darwinian picture can easily explain such trends: modest change produced by selection over a brief interval adds up to great amounts of change as geological time goes by. Nor is it difficult to imagine natural selection acting to increase brain size through time: even if brain size is only loosely correlated with degree of intelligence, there can be little doubt that our mental capacities far outstrips ape abilities. And, judging from the archeological record of the growth of technology (just to take one single and convenient measure of human intelligence and cultural complexity), there has no doubt been an increase in intelligence within the human lineage over the past 4 million years.

The picture of evolution under punctuated equilibria is, as we have seen, rather different. If species are generally stable once they first evolve, and if most evolutionary change occurs in conjunction with an actual speciation event (where a new species splits off from an ancestral species), there is no necessary accumulation of anatomical change in any particular, specific direction. Brain size, for example, need not always have increased in a particular speciation event, and it could even theoretically have happened that brain size might have decreased, as could have happened if overall body size decreased during a speciation event in human evolution. Recall the case of "Floresian Man"—three feet tall with a brain the size of a chimp or an australopith. This diminutive species lived almost to modern times in the Upper Pleistocene, with an anatomy that some think archaic human, and others see as possibly even our own species. I am pretty sure it is archaic and represents a case of island dwarfing: body size is reduced, and with it brain size.

In other words, repeated speciation events would not be predicted to yield evolutionary change in any particular single direction. Unlike Darwinian gradualism, punctuated equilibria predicts that directional trends will not necessarily, or even usually, characterize the evolutionary history of a lineage. In fact, though no one has as yet actually surveyed enough fossil lineages with this question in mind, it is likely that most lineages in fact do not display conspicuous trends.

Yet some do—including the glaring example of brain size increase in human evolution. But the fossil record reveals that individual human species do tend, in fact, to show stability. *Homo erectus* maintains an average brain size of 1000 cc. for over a million years—or shows at most only a slight increase. Apparently the same holds true of the various australopithecine species as well, at least insofar as can be told. Nor do the data indicate that our own skulls have increased in cranial capacity over the last 100,000 years. Stability within species seems to be the watchword for brain size, which is nonetheless one of the very conspicuous features that changes the most over the last six to seven million years of human evolution! How do we reconcile the conflict between the picture of stasis and speciation, on the one hand, and the undoubted long-term accumulation of directional change—this trend of increasing brain size in human evolution—on the other hand? The brains of Neanderthals were, on average, a bit larger than our own—though the expanded forebrain of *Homo sapiens* implies greater cognitive capacity in our own species than in Neanderthals.

In our paper on punctuated equilibria in 1972, Gould and I proposed that evolutionary trends may occur through a large-scale process, analogous to natural selection, but working directly on the level of species. Natural selection works on variation among individual organisms within a species. Relative success in making a living biases an organism's reproductive success, so that the more flourishing individuals will, on average, tend to leave more offspring than their less successful cohorts. Gould and I simply thought that there may be analogous factors that can bias the relative success of entire species: some species may perhaps be more prone than others to extinction, or to leaving more "offspring" species behind. In particular, there may be a process that selectively favors species with some traits over others: larger brains, for example. Species selection (to use the term coined by paleontologist Steven Stanley in 1975), in contrast, would work on the variation between species within an evolutionary lineage. The idea is that there may well be factors that bias the survival of different species preferentially: some species may be more adept at surviving than others.

Species selection is one of the additional theoretical notions that follows as an implication from the general idea of punctuated equilibria. Its precise definition, and the way it actually works, remain subjects of continuing debate within evolutionary biology. But we have already encountered enough examples of evolutionary pattern to be confident that closely related species do indeed show patterns of differential fates as time goes by. Species are sorted: some become extinct while others live on. Some tend to give rise to descendant species while others may not. There must be factors biasing the births and deaths of species—and some may well be acting to bias species survival in such a way as to produce a linear, directional trend in the evolutionary history of a lineage.

Returning to the human fossil record, let us suppose that natural selection may well act, under the appropriate conditions, to increase brain size. Yet brain size seems to have been stable over the vast majority of time that any of our predecessor species was in existence. The combination of physical and cultural adaptations of these species successfully met the environmental conditions, and stasis was the general rule despite great fluctuations in climate successfully met, for example, by *Homo ergaster/erectus* during its long 1.5 million year sojourn on Earth.

Yet, just as clearly, speciation occasionally produced descendant species with larger cranial capacities—presumably meaning with greater intelligence and consequent greater range of cultural capabilities. Modern humans are curious mixtures of specialists and generalists. We are "specialized" in relatively few features, but they are critical: our hands have opposable thumbs for manipulation; we are bipedal. And, in conjunction with our specialized stereoscopic vision and mental powers, we have enormous capabilities: culture has become our basic adaptive approach to the environment, easily transcending our primitive heritage as simple hunter-gatherers. Physically, we otherwise remain rather generalized; we are, for example, omnivores. And, it turns out, almost ironically, our culture is an evolutionary specialization which conveys tremendous flexibility when approaching the environment: our evolutionary specializations actually enhance our ecological generality. [Plates 72]

PLATE 72

***HOMO ERECTUS* AND TOOL.**

Human culture, symbolized here by its archeological traces, is an evolutionary specialization that conveys great ecological flexibility to our species. The style of this handaxe was typical of Homo erectus through much of its temporal range. The particular style of this stone tool, however, appears later in the archeological record than this earlier specimen of *H. erectus*, dated at 1.6 million years (See Plate 56).

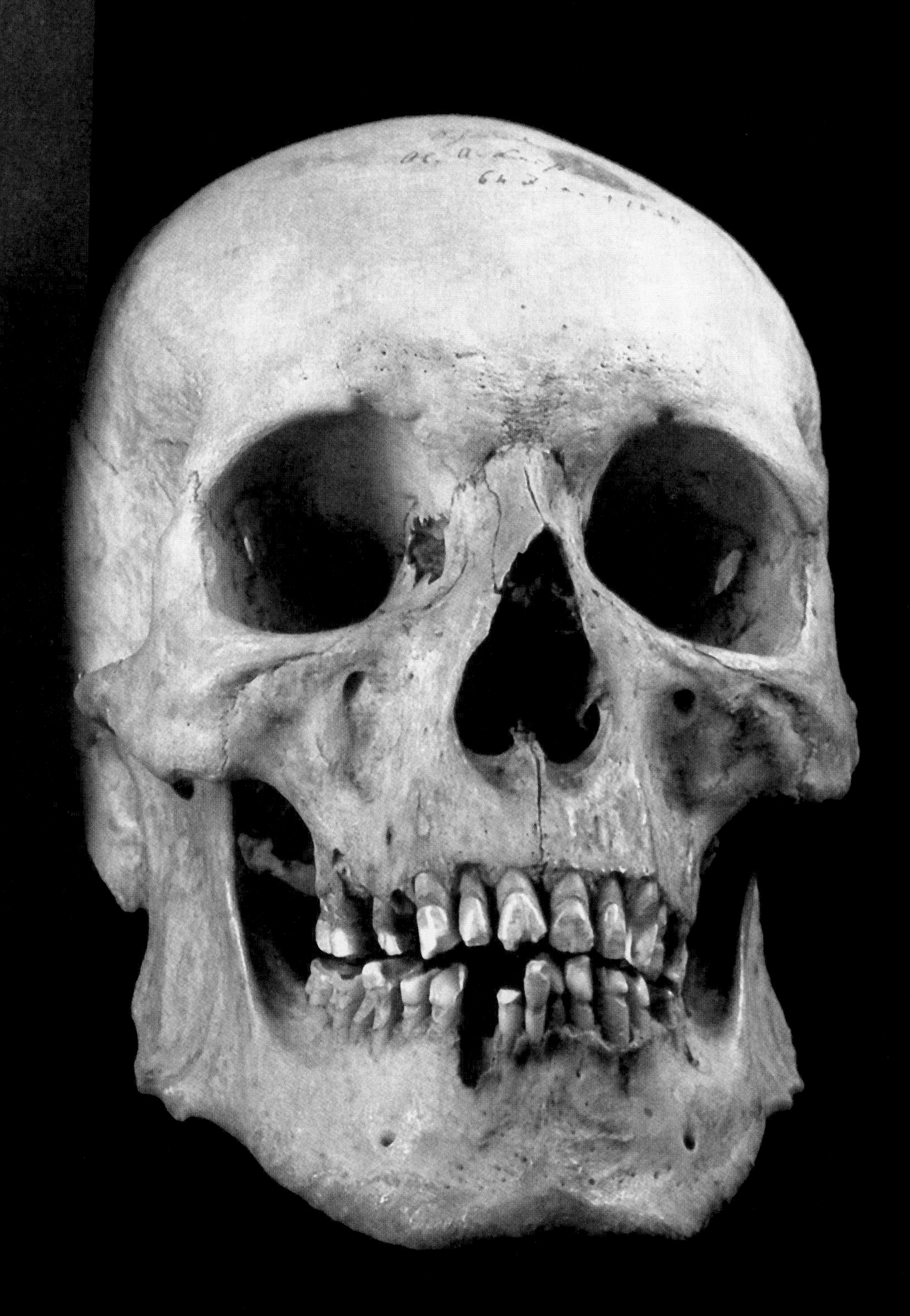

PLATE 73

Closely related ecological generalists seldom coexist side-by-side (as we shall see in more detail in Chapter 5). The only time in early human prehistory that two species of our lineage did in fact coexist, one at least (one of the several species of robust australopithecines) was an ecological specialist, bound by its specialized teeth to be a herbivore. And, though we seem to have persisted alongside Neanderthals for a time in the Near East, our arrival coincides with their demise a bit later on in Europe. The bottom line is that generalists tend to respond to competition by refusing to share the ecological pie: one generalist will always drive another (closely related) generalist out, as both are focusing on the same broad range of resources. Specialists focus on different resources, and are therefore much more likely to lead lives of peaceful coexistence.

Yet adaptive change tends to occur in speciation; and fledgling species that are ecologically distinct from parental species have a better chance of gaining an ecological foothold—and thus surviving. Recall that natural selection may well favor increase in brain size, improving intelligence and cultural capacity. I believe that the human evolutionary story, much like any other, involved a series of speciations. When such speciation was accompanied by adaptive change, predominantly a matter of enhanced cultural capacity reflected in increased brain size, fledgling species had high probability of survival. They were, if necessary, even able to out-compete their parental species.

Speciation acted like a ratchet in human evolution—enabling natural selection to increase brain size in isolated populations. New species, equipped with superior cultural capacities, the locus of our primary ecological adaptation, were bound to succeed. And thus, by leaps and bounds, did brain size increase, as predicted, in human evolution. [Plate 73]

PLATE 73

***HOMO SAPIENS,* MODERN SKULL, WITH A COPY OF THE ORIGINAL EDITION OF *THE DESCENT OF MAN*, BY CHARLES DARWIN.**

Darwin, in his *On the Origin of Species,* dared only to suggest that his insights on evolution would "shed light" on the origins of our own species. In *The Descent of Man*, he completed his thesis, connecting *Homo sapiens* fully and explicitly with the evolutionary history of all other living things.This early human ancestor lived very close to the time of the chimpanzee-human divergence.

CHAPTER 5

LIVING FOSSILS

PLATE 74

***LATIMERIA MENADOENSIS*, AN INDONESIAN COELACANTH.**

Recent. Manado Tua Island, Sulawesi, Indonesia.

In 1938 a fisherman landed a coelacanth, a fish previously known only from 66-million-years-old fossils, in the Indian Ocean off the coast of the Comoros Islands. In 1997 a second species of coelacanth was discovered over 5,000 miles away in the Pacific Ocean in waters off the Indonesian island of Manado Tua.

Credit: Mark Erdmann, Conservation International

PLATE 75

UNDINA PENKILLATA,
COELACANTH FISH.

Upper Jurassic (*ca.* 150 million years). Solnhofen Limestone, Bavaria, Germany.

Coelacanths first arose in the Devonian, and continued with little change throughout the Mesozoic.

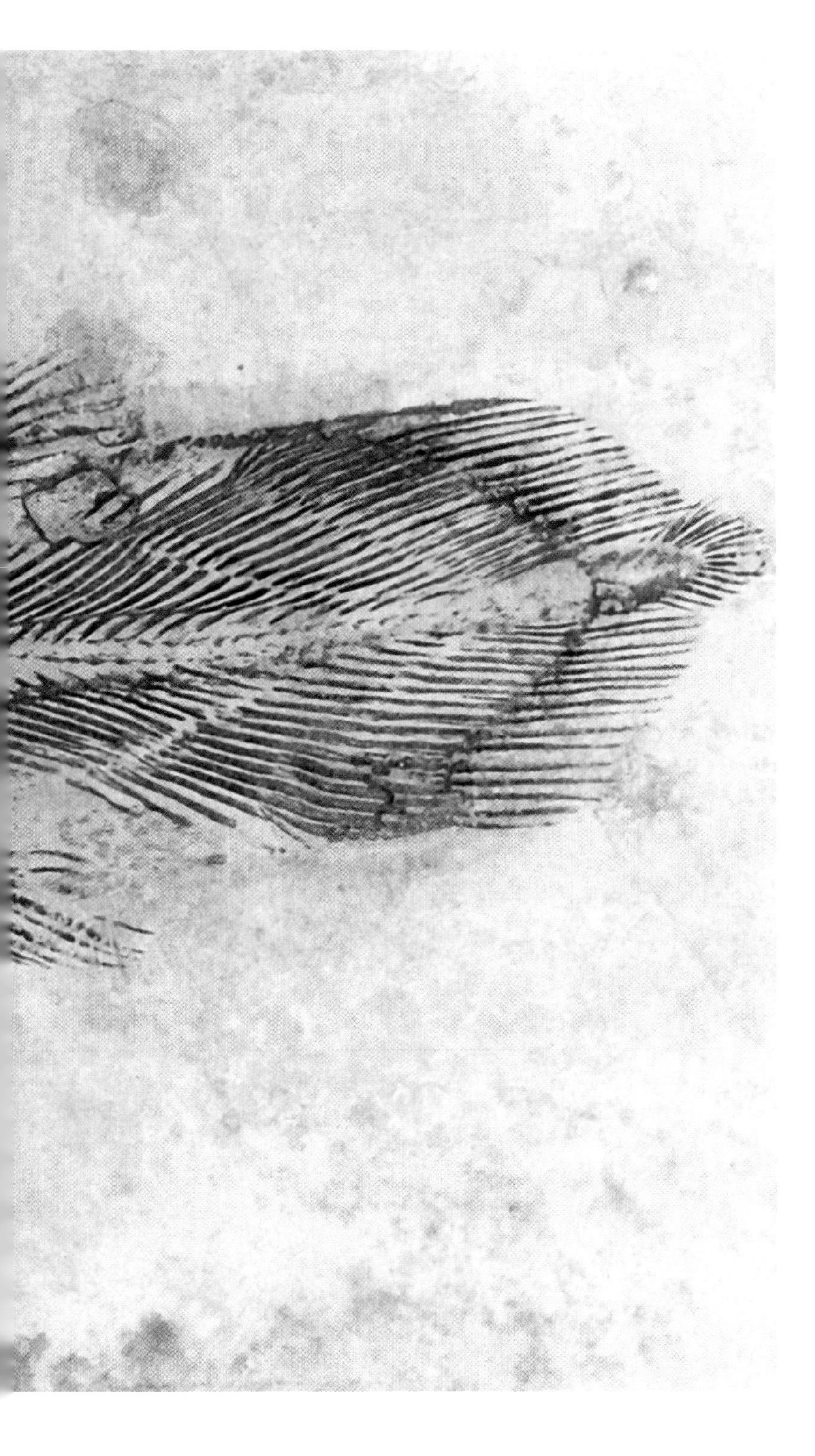

Every so often, newspapers proclaim the discovery of an important new fossil. We may read of further finds of human fossils in Africa, or of additional unusual dinosaurs that keep turning up around the world. These are the most striking (and newsworthy) of a virtually unbroken stream of new species of all kinds that paleontologists are continually unearthing.

We know a lot more about species still living on Earth than those that have lready become extinct. Yet occasional surprises turn up in the modern world, too. In 1938, a fisherman off the coast of South Africa landed a large, dark scaly fish unlike any other he'd seen before. Odd encounters such as these mostly pass unnoticed by the world of professional biology. But this particular fisherman was, thankfully, curious about his catch; in due course, the fish came to the attention of Marjorie Courtenay-Latimer, a curator at one of the local museums. She realized that the fish resembled nothing yet seen in the modern fish fauna. But it did look hauntingly similar to some illustrations of extinct fossil fishes. Turning to the literature, it quickly became obvious that the fish in question (in due course dubbed *Latimeria chalumnae*) was nothing less than a living member of the coelacanths–a group that had presumably become extinct some 66 million years ago! [Plate 74, see page 120.]

The find galvanized zoologists and paleontologists alike. Coelacanths had had a very long evolutionary history. Arising in the Middle Devonian. they flourished throughout the Upper Paleozoic, sailed through the great extinction that divides Paleozoic from Mesozoic times, radiated into a diverse array in the Triassic Period and lived on through to the end of the Cretaceous. [Plate 75] But, insofar as anyone knew, they had not survived the grand extinction event that took out the dinosaurs and so many other forms of Mesozoic life. The latest coelacanth yet found in the fossil record is the 70 million year old *Cretaceous Macropoma*, in many ways the spitting image of Latimeria, though at less than a foot in length, only a fraction as long as its four-foot living relative. [Plate 76]

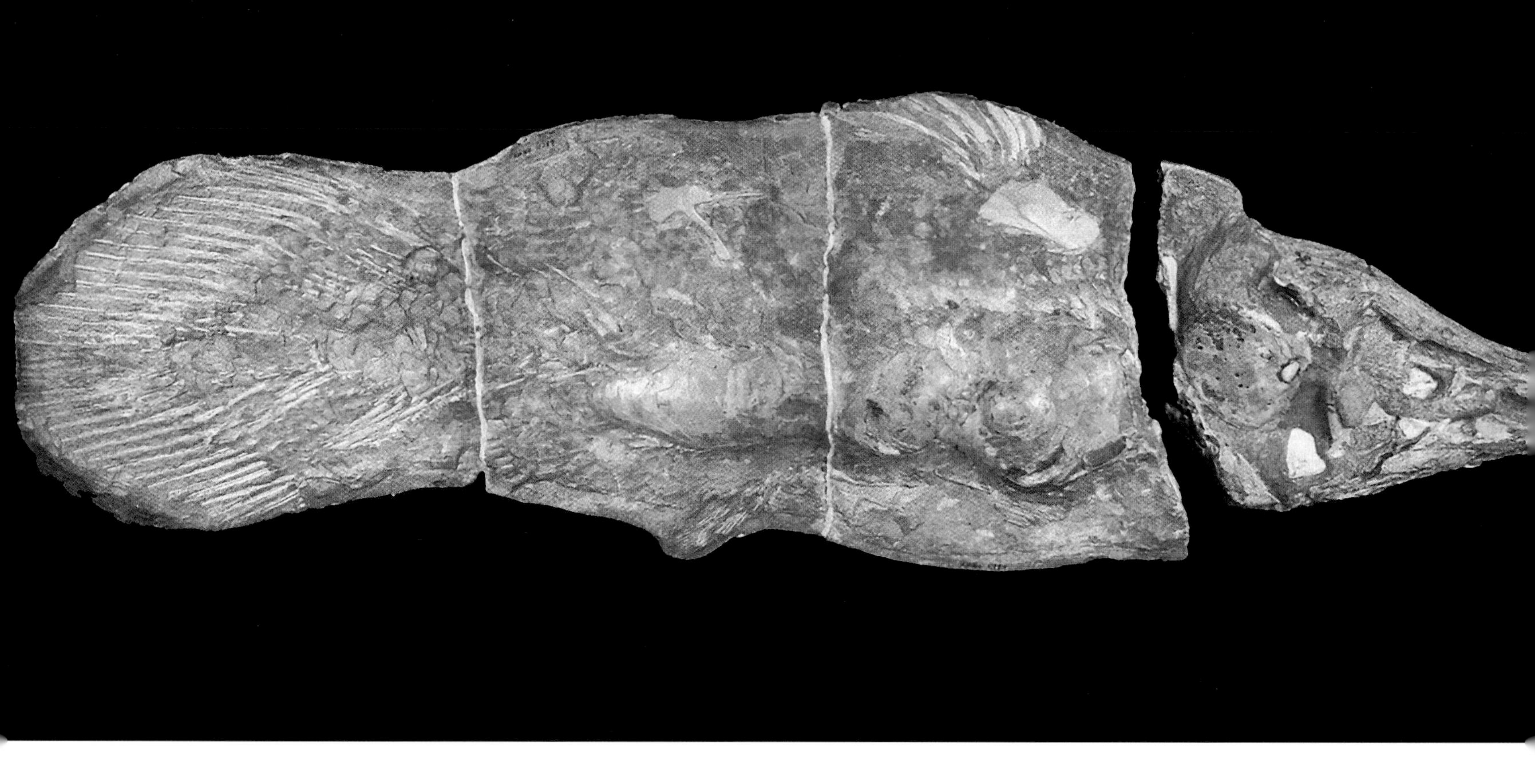

PLATE 76

***AXELRODICHTHYS ARARIPENSIS,* COELACANTH FISH.**

Lower Cretaceous (*ca.* 115 million years). Brazil

Coelacanths disappear from the fossil record after the Cretaceous leaving a gap of over 66 million years between the geologically youngest fossil and the modern *Latimeria.*

PLATE 77

OLENEOTHYRIS HARLANI
TEREBRATULID BRACHIOPOD, OR LAMPSHELL.

Middle Paleocene (*ca.* 60 million years). New Egypt, New Jersey

This species closely resembles members of one of the only two remaining groups of toothhinged ("articulate") brachiopods in the modern seas.

Coelacanths are lobe-finned fishes. Their closest relatives are the extinct rhipidistians, still most paleontologists' candidates as nearest relatives to the amphibians that managed to crawl out onto land sometime in the Middle or Upper Devonian. Other zoologists opt for lungfishes as the closest relatives of terrestrial vertebrates. Lungfish, like coelacanths, are still alive today, with a species each in Australia, South America and Africa. Like the coelacanths, the first lungfish also appeared in the Devonian, approximately 385 million years ago. By Upper Devonian times, there was rather a wide assortment of different forms, including the rakish, actively predaceous Fleurantia, which looked far more like a rapacious mackerel than a modern lungfish. After the Devonian, lungfish were greatly reduced in numbers. Yet they managed to hang on right up to modern times.

But lungfish lack some of the dramatic glamour that the discovery of *Latimeria* held: modern lungfish were known before their fossil remains were discovered and studied to any great detail. Coelacanths, in contrast, had been known strictly from fossil remains prior to the unexpected and happy fisherman's catch. No others have ever appeared again–off the coast, that is, of southern Africa, where that first one turned up. Fortunately, zoologists eventually learned that fishermen catch *Latimeria* at irregular intervals off the Comoro Islands 1,500 miles to the northeast of Africa and just off the northwest coast of Madagascar in the Indian Ocean. Now, fishermen have become so adept at catching the highly desirable *Latimeria* that the species is reported to be under the imminent threat of extinction. What an irony for the sole surviving species of a lineage that managed to hang on for so long!

Latimeria and lungfish are but two of the more famous cases of "living fossils."*Latimeria*, in particular, seems literally as if fossils had suddenly "come alive." The term "living fossil" actually embraces a wide number of cases and has no hard and fast definition. But, in general, a living fossil bears a strong resemblance to its fossil forebears. [Plate 77] They have a Mesozoic or even a Paleozoic "leer": they tend to look almost hauntingly archaic. They look as if evolutionary time has stood still, as if they more appropriately belong to a bygone era in the remote geological past.

In general we should not be surprised that ancient structures are conserved. [Plates 78, 79] Evolution is not a simple game of progress, with newly evolved forms forever driving out the old. [Plates 80, 81] Such may happen in some lineages; as we have seen in Chapter 4, much of human evolution can be interpreted as a matter of bigger-brained species surviving, while their less well-endowed ancestral kin die out. Bacteria, however, were neither displaced nor replaced by eukaryotic algae and other, more animal-like single celled forms. Nor did the rise of multi-celled animals and plants displace the inhabitants of the microbial world. Moreover, evolution inevitably entails the modification of some structures at a faster rate than others: bipedalism, as we have seen, came early in human evolution, while brain size increase accumulated at a far more leisurely rate. And, in many ways, we have changed little if any beyond the condition of the ancestor we held in common with apes: it is a telling statistic, indeed, that we share 98.5% of our genetic information with chimpanzees.

PLATE 78

NUCULOIDEA DECEPTRIFORMIS,
A PROTOBRANCH BIVALVE MOLLUSK, OR "NUT CLAM."

Middle Devonian (*ca*. 380 million years). Morrisville, New York

Protobranchs are primitive bivalves.

PLATE 79

PLATE 79

ENNUCULA SUPERBA,
A NUT CLAM.

Recent. Budaberg, Australia.

Nut clams have changed very little since the Paleozoic.

PLATE 80

LINGULA ANATINA,
A TOOTHLESSHINGED OR "INARTICULATE" BRACHIOPOD.

Recent, The Phillipines.

The animal is able to pull itself down into the seafloor by contracting its long, fleshy pedicle.

PLATE 81

That was Darwin's very argument that evolution must have occurred: "descent with modification" will produce a nested pattern of similarity linking up absolutely all living creatures. The closer two species are related, the more similarities they will share. But this implies that there must be some similarities that are shared by absolutely all living beings. And this is so: all organisms have the information molecule RNA, so important in transcribing DNA and regulating the synthesis of proteins. There is a part of the gene sequence of virtually all organisms known as "ubiquitin." The structure of the cells of our mammalian bodies is similar in all essentials to the structure of rose plant cells. All mammals have hair. Once a feature appears in evolutionary history, it may well not change much for eons—and so be passed on to all later descendants. But these are *parts* of organisms.

The classic cases of "living fossils" reveal a more pervasive conservatism: there seems to have been almost no change in any part we can compare between the living organism and its fossilized progenitors of the remote geological past. Living fossils embody the theme of evolutionary stability to an extreme degree. [Plates 82, 84, 85] They are at one far end of a spectrum of rates and degrees of evolutionary change. Against them we might pit the mutability, the evolutionary changeability, of disease-causing and antibiotic-resistant *staphylococcus* bacteria, malaria pathogens or the dreaded retroviruses (that cause AIDS and other horrid afflictions): in the short term, at least, evolutionary change in these microbes has been extremely rapid. And so we ask: what underlies this great disparity of evolutionary rates? How can some lineages remain nearly unchanged through tens and even hundreds of millions of years—while most others, evolving a lot more rapidly, accumulate a great deal of evolutionary change?

PLATE 81

LINGULA QUADRATA,
INARTICULATE BRACHIOPOD.

Middle Ordovician (*ca.* 460 million years). Trenton Falls, New York.

Details of muscle insertion sides on the interior sides of inarticulate brachiopods shells demonstrate that modern *Lingula* have diverged but little from their Paleozoic ancestors.

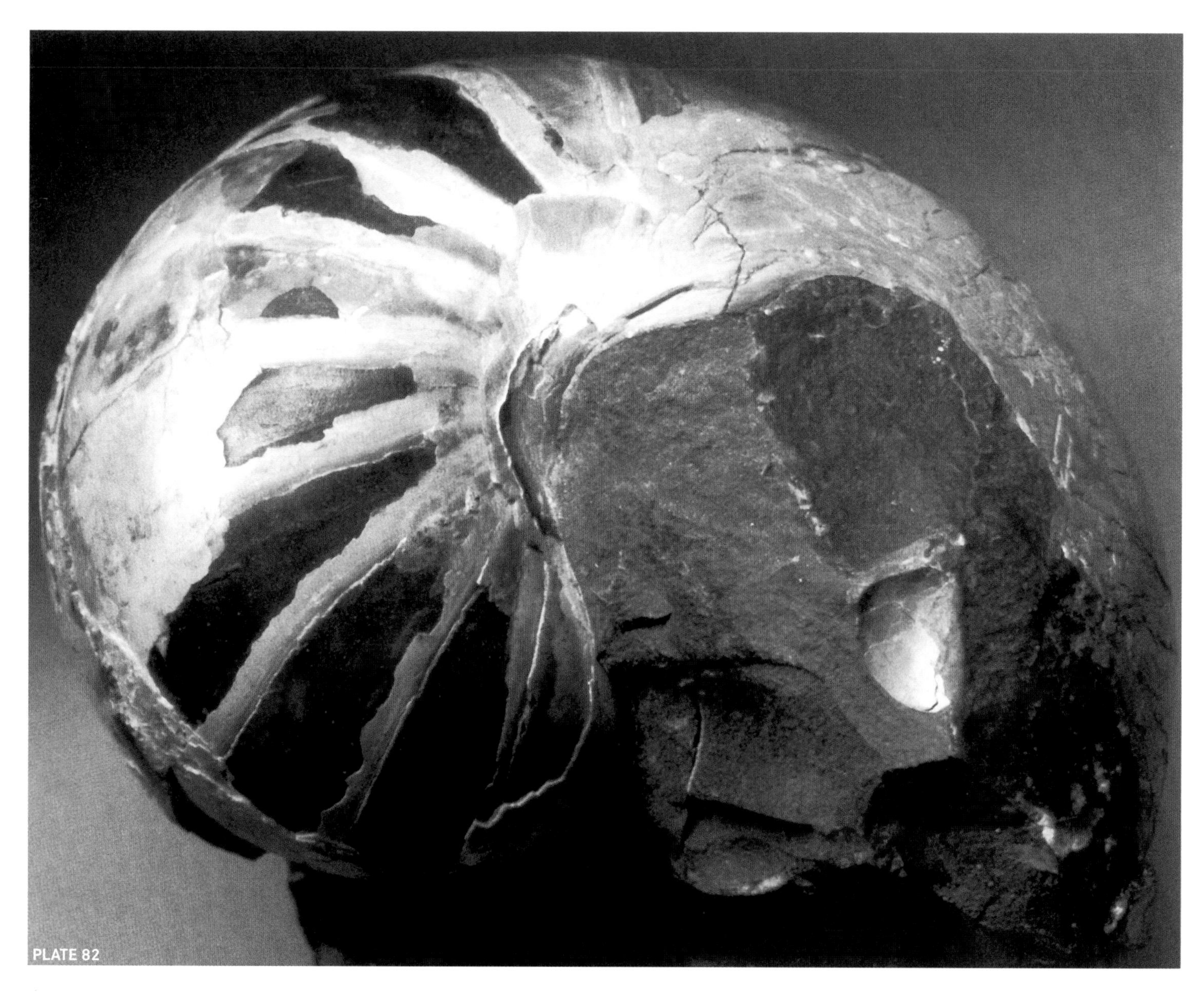

PLATE 82

EUTREPBOCERAS DEKAYI,
NAUTILOID CEPHALOPOD MOLLUSK.

Upper Cretaceous (*ca.* 80 million years). South Dakota.

This specimen reveals the internal partitions of the shell, in which the original nacre ("mother-of-pearl") is exquisitely preserved.

PLATE 83

NAUTILUS POMPILIUS,
NAUTILOID CEPHALOPOD MOLLUSK.

Recent. Western Pacific.

A classic living fossil, living *Nautilus* closely resembles its Cenozoic, Mesozoic–and even its Paleozoic–predecessors.

PLATE 83

PLATE 84

TRYBLIDIUM OVATA,
MONOPLACOPHORAN MOLLUSK. CAST AND MOLD REVEALING MUSCLE SCARS OF INTERIOR OF SHELL.

Lower Ordovician (*ca.* 480 million years). Fort Cassin, Vermont.

Monoplacophora, the most primitive class of the Phylum Mollusca, were thought to have become extinct by the end of the Middle Devonian. Modern *Neopilina* species, however, were dredged from the deep oceans in the 1950s; they closely resemble this ancient species from the Lower Ordovician of Vermont.

PLATE 85

LIMULUS POLYPHEMUS,
HORSESHOE CRABS.

Recent. Nantucket, Massachusetts.

One of four living species of horseshoe crabs, this species occurs along the Atlantic coast from Nova Scotia south to the Yucatan Peninsula.

Credit: Chosovi.

ADAPTATION AND LIVING FOSSILS

Horseshoe crabs are classic living fossils. They are the last surviving members of a once-diverse array of Paleozoic denizens of the seas. Not true crabs at all, horseshoe crabs are actually distant kin of spiders, mites, ticks and scorpions—a distinctly terrestrial cadre of arthropods. Horseshoe crabs, it was realized towards the end of the nineteenth century, lack the antennae typical of insects and crustaceans (true crabs, plus lobsters, barnacles and a host of other, predominantly though by no means exclusively, marine creatures). In that they resemble spiders and scorpions. The clincher, though, is that *Limulus polyphemus* (the horseshoe crab of eastern North America) and its relatives in Asian waters have a pair of pincers ("chelicerae") around the mouth—just as all ticks, mites, spiders and scorpions do. [Plate 85] Chelicerae virtually define the Class Chelicerata: their presence in horseshoe crabs tells us that these beasts are by no means true crabs, but belong instead with spiders and their relatives. They also tell us where to look for the ancestry of the terrestrial scorpions and spiders: in the marine deposits left behind by Middle Paleozoic near-shore seaways. There are four living species of horseshoe crab. Curiously enough, all live along the eastern margins of the two great northern hemisphere continental landmasses of Eurasia and North America. Truth to tell, there isn't a whole lot of difference between these four species.

Horseshoe crabs have had a rich and varied history. Like lungfishes, they were particularly diverse back in the Paleozoic, not too long after they first evolved. Some years ago, I was fortunate enough to have a spectacularly well-preserved early horseshoe crab show up in my office. The specimen was brought in by the Reverend LeGrand Smith, an expatriate American school administrator who had been living and working in the vicinity of La Paz, Bolivia. LeGrand is an ardent amateur paleontologist. I cannot overemphasize the importance that the thousands of amateur paleontologists like LeGrand Smith continue to have in influencing the course of intellectual history. Fossils are very much like berries: berries ripen in the fall, gustatory targets for birds and mammals, certainly including ourselves. Evolution has fashioned berries to be attractive so that seeds will be dispersed away from the parental plant. If someone doesn't eat them, they will rot on the vine. Fossils weather out by the billions and, like berries, disintegrate and "rot" away if not picked up by someone who cares, someone who appreciates their potential real significance.

The specimen Smith brought in to the American Museum of Natural History back in the early 1970s was very primitive, even by horseshoe crab standards. [Plate 86] Modern horseshoe crabs have a large, crescentic (horseshoe-shaped) head region (the "prosoma"), with the closest thing to "beady" little eyes imaginable in an animal with compound eyes sporting hundreds of individual lenses. There are six pairs of walking appendages under the head. The middle body portion is a fused plate, housing gills below. The last part of the body is that long and rather wicked-looking tail spike ("telson")—which, contrary to popular imagination, carries no venomous sting. The telson is used, instead, to help the animal right itself when it finds itself occasionally flipped over upside-down. With their legs flailing away helplessly, horseshoe crabs would be doomed, lying there on the seafloor on their backs, were it not for that tail spike.

In sharp contrast to modern horseshoe crabs, the head region of the Bolivian fossil has huge crescentic eyes. And, instead of that single fused plate forming the middle body region, the fossil has a series of segments—very similar to a trilobite's body, which allowed the animal to roll up into a ball in life. If you hold the shed skin of a young horseshoe crab up to the light, you can often see faint traces of these original free segments, now all fused together into a single, solid rigid

PLATE 86

***LEGRANDELLA LOMBARDII,* PRIMITIVE ARTHROPOD RELATIVE OF HORSESHOE CRABS. HORSESHOE CRABS.**

Lower Devonian (*ca.* 390 million years). Bolivia.

In this species, the segments of the body ("opisthosoma") are still free, like those of a trilobite, and unlike the fused, solid plate of an advanced horseshoe crab.

PLATE 87

***EUPROOPS DANAE,* RUBBER MOLD AND PLASTER CAST OF HORSESHOE CRAB.**

Middle Pennsylvanian (*ca.* 310 million years). Illinois.

This species was a denizen of freshwater coal swamps, and not a direct ancestor of modern horseshoe crabs.

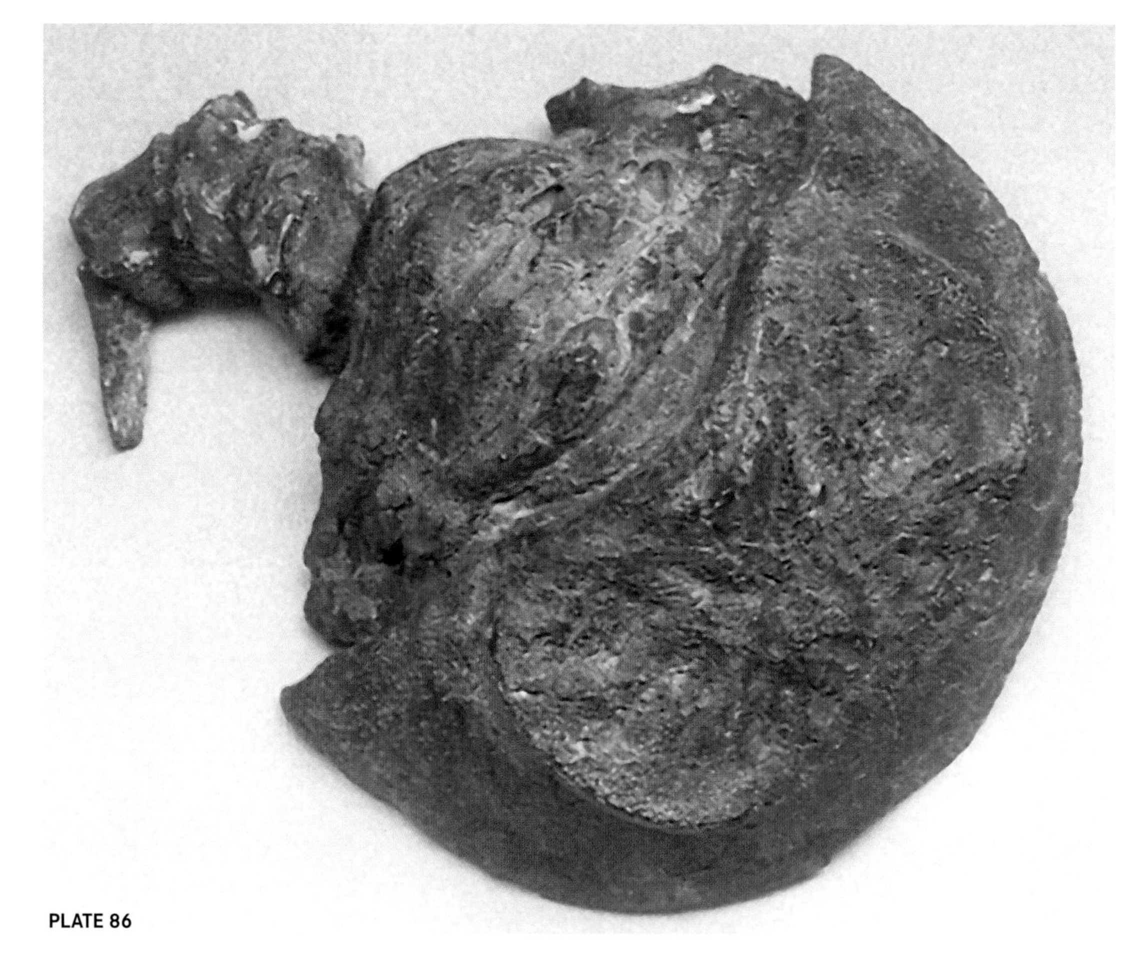

PLATE 86

plate. Yet, behind these segments of the Bolivian fossil lay the tell-tale spiky tail—looking just like its counterpart in modern horseshoe crabs.

My Bolivian fossil is Lower Devonian in age, approximately 390 million years old. (I named it, incidentally, *Legrandella lombardii*, honoring, first, its discoverer, and, second, the technician, Frank Lombardi, who so painstakingly extricated the specimen from its encasing rock matrix.) With a few other specimens we were able to garner, *Legrandella lombardii* shed much light on the details of early horseshoe crab evolution, from the dim mists of their beginnings some 450 million years ago in the Ordovician Period, up through the evolution of some strange beasts imperfectly known from Silurian and Devonian rocks of Europe and North America.

By Upper Devonian times, though, horseshoe crabs had already fused that middle region of their bodies. A new age had dawned. In the vast swamp deposits that have supplied most of the world's coal, little horseshoe crabs with notably rounded middle body regions are fairly common. Evidently accustomed to fresh water and perhaps even to withstanding life exposed to air, these little *Euproops* horseshoe crabs begin to take on the look of modern horseshoe crabs. [Plate 87] But, as it turns out, these coal swamp dwellers were yet another of life's side-branches. The real precursors of modern horseshoe crabs were living (as you might expect) in marine environments. It is with these creatures that we get the analogue of the coelacanth fish *Macropoma-Latimeria* resemblance. Pennsylvanian-aged (*ca.* 310 million year old) Paleolimulus looks very much like *Limulus*. [Plate 88] As with the two fish, the most obvious difference, at least at first glance, is simply size: modern horse crabs are many times the size of the largest of the *Paleolimulus* specimens yet collected. And that's about all there has been to horseshoe crab evolution: apart from the bizarre Triassic *Austrolimulus fletcheri*, which looks something like a flying wing [Plate 89], horseshoe crabs settled down to a humdrum, very slowly evolving course that has lasted, so far, some 300 millions of years.

Horseshoe crab behavior and ecology can help us understand the reasons for their extreme evolutionary conservatism. They are often the very last species to be driven from an estuary as human pollution transforms the environment far beyond the tolerance limits of most species. And that is precisely the point.

PLATE 87

PLATE 88

***PALEOLIMULUS* SP., A HORSESHOE CRAB FOSSIL WITH RARE PRESERVATION OF FEET.**

Middle Pennsylvanian
(*ca.* 310 million years).
Mazon Creek, Illinois.

This species is closely related to modern forms—and appears almost to be a miniature version of modern species.

Credit: The Fossil Forum.

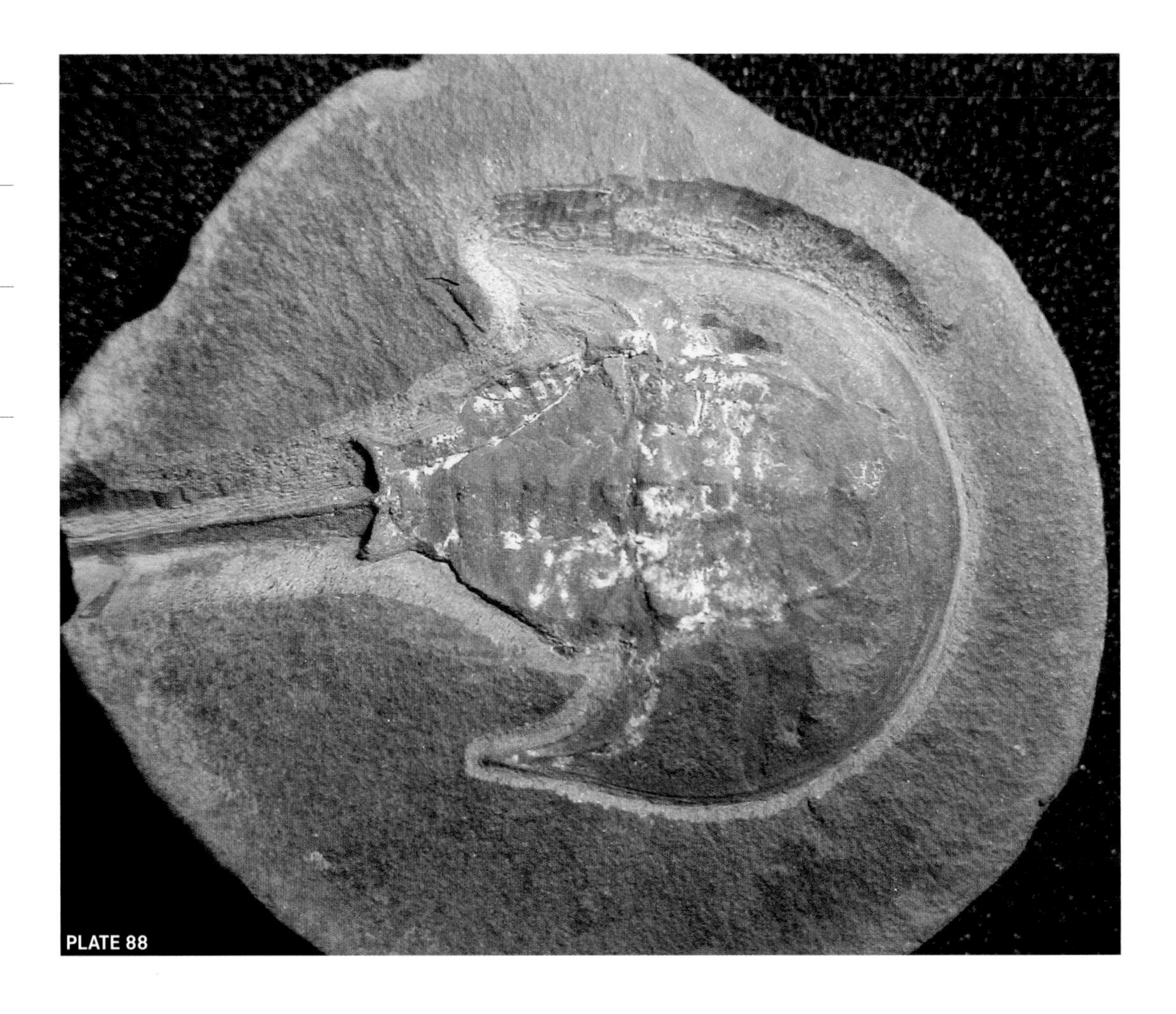

PLATE 89

***AUSTROLIMULUS FLETCHERI,* A BIZARRELY-SHAPE HORSESHOE CRAB.**

Triassic (precise age uncertain).
Near Sydney, Australia.

Austrolimulus is the only horseshoe crab known to have departed significantly from the conservative body plan typical of the lineage as a whole.

PLATE 89

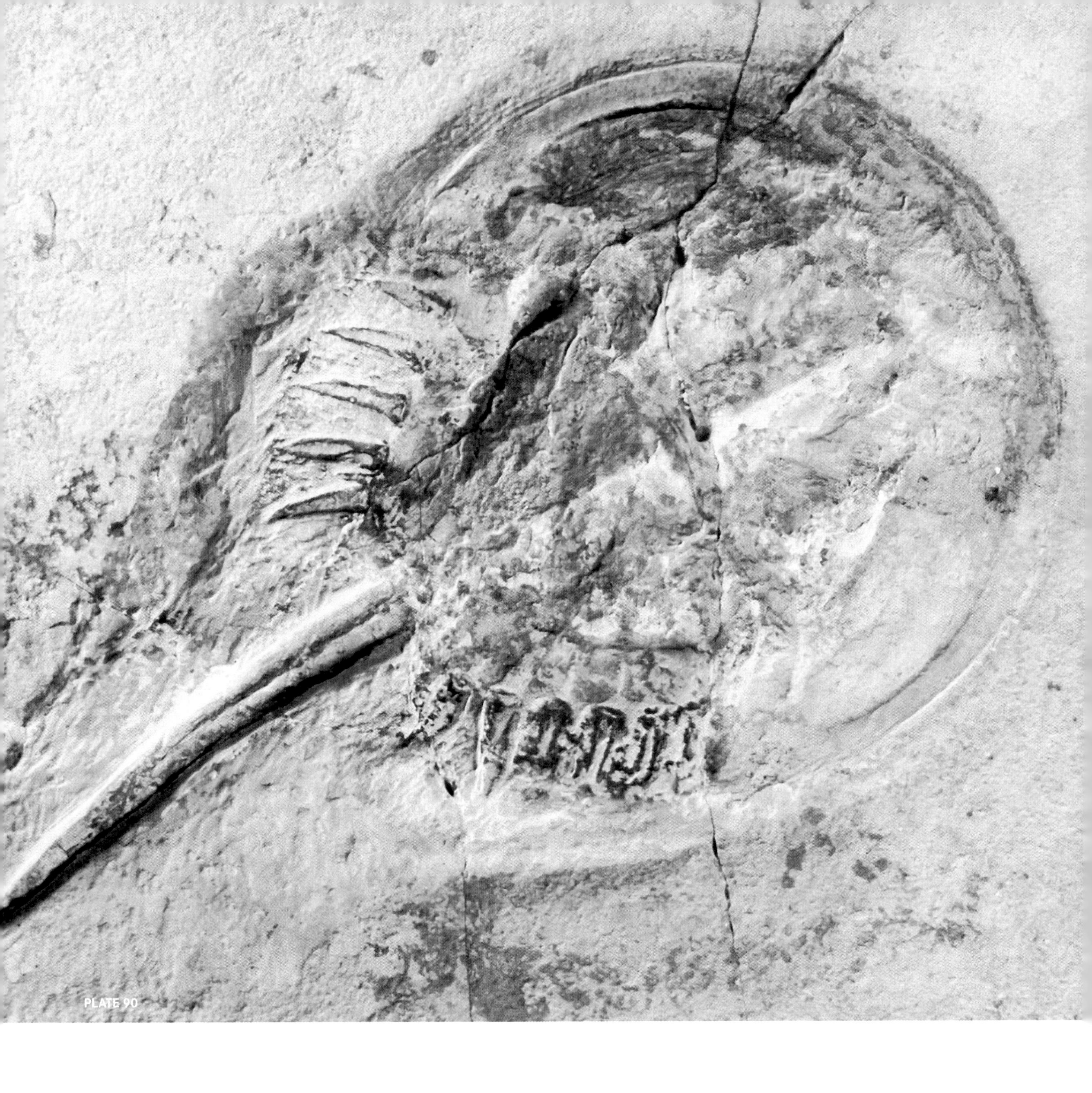
PLATE 90

Horseshoe crabs are ecological *generalists*: they can withstand a wide range of conditions, natural as well as human-induced. Marine creatures, they can tolerate wide fluctuations in salinity, both the briny conditions of heightened salt content, as well as the brackish conditions typical of estuaries, where fresh riverine waters mix in with the salty sea. Because they clamber out from the sea to lay their eggs in nests on sandy beaches, they can tolerate some exposure to air. They are so tough that they have been known to linger on land for days, trapped away from their natural watery home, exposed to the fierce rays of solar radiation. They tolerate an impressive range of temperatures and oxygen availability. And they eat a wide range of food items: they catch worms and small mollusks, crushing and chopping them up into fine bits with the bases of their legs, before sending them down the gullet. But they also scavenge, surviving quite well on bottom debris. All told, horseshoe crabs are very impressive in their hardiness. [Plate 90]

PLATE 90

MESOLIMULUS WALCHI,
BOTTOM VIEW OF A MESOZOIC HORSESHOE CRAB.

Upper Jurassic (*ca.* 150 million years). Solnhofen Limestone, Bavaria, Germany.

Detailed studies have shown that, despite its obvious similarities with modern species, *Mesolimulus* was flatter, and was capable of swimming a bit faster than modern species.

And that hardiness, that jack-of-all-trades nature that stamps horseshoe crabs so clearly, holds the clue to why they have remained so stable, so evolutionarily non-changeable for hundreds of millions of years, as we have noted in Chapter 4. Ecologically generalized species do not seem very prone to sharing the resources of their habitats with close relatives. Usually, when closely related species coexist in the same region, they will differ in how they approach their surroundings. The effect is especially noticeable in the search for food in animals: foxes, coyotes and even wolves may live in the same general region, but each species will be concentrating on different food items. Even though coyotes and foxes may overlap in their dining preferences, the bulk of the diet of each species will be different: foxes will take more field mice, while coyotes will be more successful at bagging rabbits. These species are somewhat specialized, and avoid competition by not going after exactly the same menu items. [Plate 91]

PLATE 91

***GERATONEPHILA BURMANICA*, ORBWEAVER SPIDER ATTACKS PARASITIC WASP, *CASCOSCELIO INCASSUS*.**

Lower Cretaceous (*ca.* 140 million years). Hukawng Valley, Myanmar.

A rare example of a spider with its wasp prey entrapped in dinosaur age amber.

Credit: George Poinar, Jr., Oregon State University.

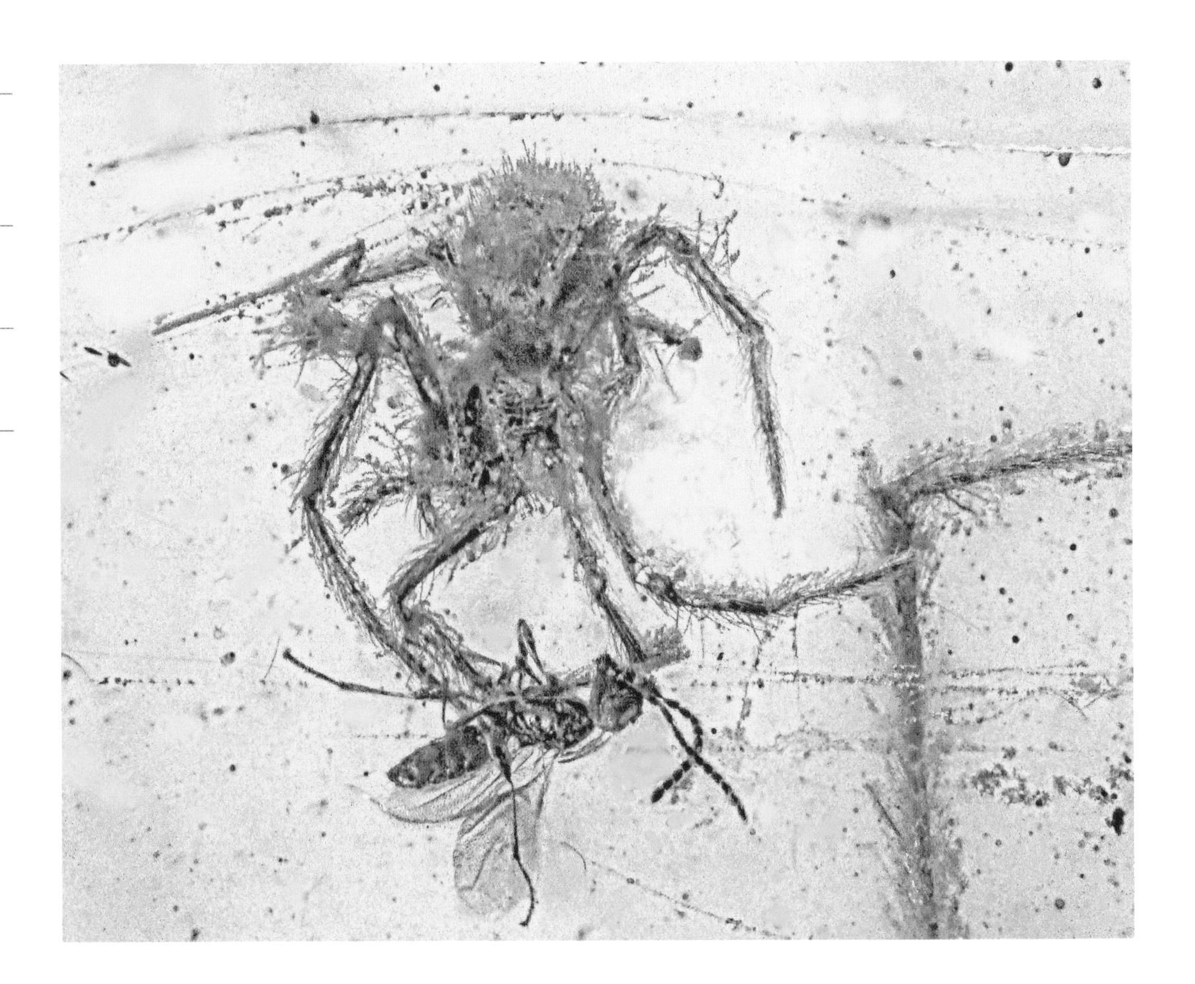

Such is not the case with generalized feeders, species that are happy to take virtually anything that comes their way. Baby horseshoe crabs may lack the strength to crack the larger clams—but their elders can handle a great range of size and manage to eat virtually any invertebrate they find lying on or just below the surface of the sea bottom. If there is a species with such a varied, eclectic diet, there really is little room for two. A closely related and ecologically similar species is likely to have a difficult time establishing a foothold. There is no inclination to "share" on the part of ecologically generalized species.

Paleontologists have realized for over a century that ecologically generalized species tend to occupy broad regions, to be found in a relatively wider range of environments and, tellingly, to last longer than their more specialized counterparts. There is a direct lesson in extinction here: the more flexible a species is ecologically, the greater the chances that it will be able to find suitable habitats as the environment changes through time. As we have seen, this is the very basis of survival—and survival without much noticeable evolutionary change.

Further, the more specialized a species is, the less likely that it will be able to continue to recognize suitable habitat as conditions change. Most vulnerable are species focusing on single food sources. Parasites often have only a single host species—which is fine, so long as the host survives. Taking advantage of an abundant single resource makes good ecological and evolutionary "sense." But such a strategy is effective only as long as that resource remains available. The trade-off is vulnerability to extinction, and it is a simple fact of nature that ecologically specialized species become extinct at much higher rates than do ecologically generalized species. There is a balance here: natural selection will tend to produce ecologically specialized organisms, as often they will flourish more than generalists in the short run. But extinction keeps pruning back the ranks of the specialists. In the long run, the generalists hang on, while the ranks of specialists are rapidly replenished by the continual evolution of new species.

Yet we have not completely solved the riddle of living fossils. After all, no one says that the modern American horseshoe crab species *Limulus polyphemus* itself goes way back to the Devonian. In fact, the leathery skeleton of a horseshoe crab does not lend itself terribly well to fossilization, and no fossilized specimens of the actual species *L. polyphemus* have as yet been found (though a few specimens of a closely similar species of *Limulus* are known from Tertiary rocks). The fact that ecological jacks-of-all-trades tend to avoid extinction on average better than specialized species is only part of the answer. Living fossils, because they are usually ecological generalists, belong to lineages that typically have very few species alive at any one time. New species do not seem to appear as often as they do in lineages formed by specialist species.

Here, again, ecological behavior tells us why speciation occurs a lot less frequently among generalists than specialists. That selfsame propensity to exclude closely-related competing species reduces the chances for new, offspring species to gain an ecological foothold. New species have a better chance of surviving if they are ecologically a bit distinct from their parental species. Fledgling species derived from generalists will tend to be rather generalized, too. They stand little chance of survival; most often, they will simply not gain a separate ecological foothold. [Plate 92]

PLATE 92

***METACRYPHAEUS TUBERCULATUS,* PART AND COUNTERPART OF A TRILOBITE ARTHROPOD.**

Lower Devonian (*ca.* 390 million years). Bolivia.

This common species lies close to the root stock of a group of rapidly evolving, and presumably ecologically specialized trilobites—three of which are illustrated in plates 93, 94 and 96.

Offspring of species already somewhat specialized seem to have a greater chance of concentrating on a particular portion of the resources not completely exploited by the parental species. New species commonly form near the edge of their ancestral species' range. The habitat at the edge is not optimal for the parental species, a situation that offers an opportunity to the fledgling species if it can adjust to the conditions at the edge of the range—experiencing the environment, which was marginal for the parental species, as optimal for itself. It is a fact of nature that specialists tend to speciate more and, often, to accumulate ever more specialization. Though they tend to be in the longer run more prone to extinction, specialists show much higher rates of speciation—simply because new species arising from specialist parents have a greater chance of surviving in the short term. [Plate 93] There is a ratchet-like mechanism of the rapid accumulation of evolutionary change as lineages keep splitting and new species bud off from old ones within specialized lineages. Without a comparable degree of successful speciation, generalist lineages have far fewer species alive at any one time than their specialized counterparts. And that means that generalists will appear to evolve at far slower rates: they are simply not generating many new species, and the few that manage to evolve show little change over their ancestors.

PLATE 93

MALVINELLA BUDDAE,
A TRILOBITE.

Lower Devonian (*ca.* 390 million years). Sica Sica, Andes, Bolivia.

This species is quite spiny and advanced over the preceding species. It lends its name to the rapidly evolving "*Malvinella*" group of anatomically advanced and specialized trilobites, including the subsequent two species.

Credit: Fossil Pictures.

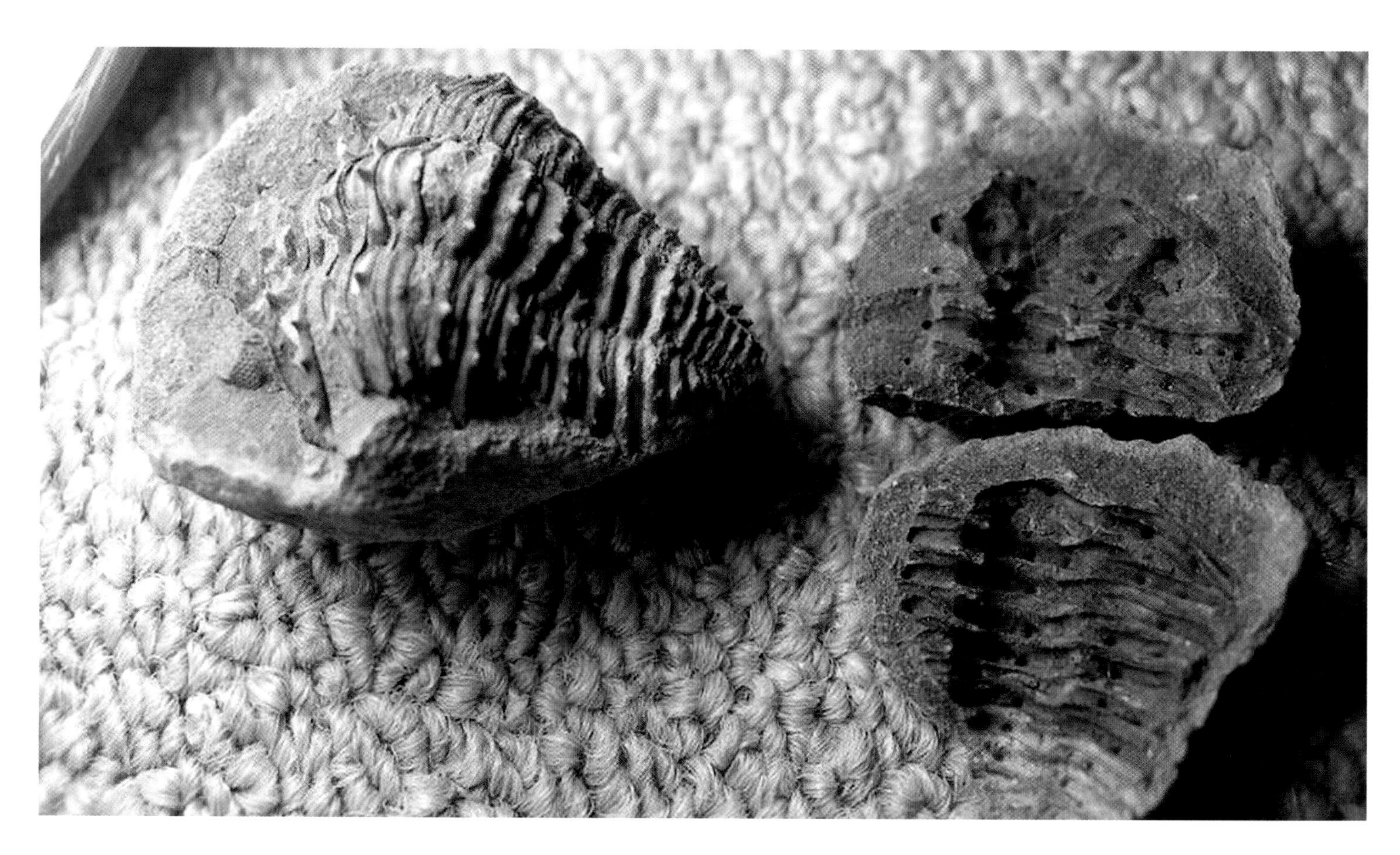

PLATE 94

***VOGESINA SP.*, A TRILOBITE.**

Lower Devonian (*ca.* 390 million years). Bolivia.

This species lost all its spines and head furrows—possibly reflecting the adoption of a burrowing mode of life.

PLATE 95

MONGOLARACHNE JURASSICA,
A WEBWEAVING SPIDER.

Middle Jurassic (*ca.* 165 million years). Daohugou, Inner Mongolia, People's Republic of China.

Because spiders and other terrestrial invertebrates are so rarely preserved in the fossil record, the pulse and pace of their evolution is not well understood. It is theorized the rise of this and other sophisticated aerial web spinners was in response to the rapid diversification of insect groups during the Jurassic.

Credit: Paul Selden, University of Kansas.

PLATE 96

***BOULEIA DAGINCOURT,* A TRILOBITE.**

Lower Devonian (*ca.* 390 million years). Sica Sica, Andes, Bolivia.

Still another specialized member of the *Malvinella* group, this trilobite evolved so far as to actually converge on *Phacops*-like species.

And so it can go on through millions of years. In extreme examples, it can go on for hundreds of millions of years—yielding our classic cases of living fossils. *Lingula* is a genus of primitive brachiopods; modern species look for all the world like their ancient Ordovician counterparts living 450 million years ago. Whooping cranes, now on the verge of extinction, belong to an ancient lineage of birds that has not accumulated much change since first appearing back in the Eocene, some 50 million years ago.

But for every astounding case of evolutionary conservatism, there are far more cases of rapid diversification and accumulation of a great deal of evolutionary change. These "adaptive radiations" are really the flip-side of the living fossil story. One of the last great episodes of rapid diversification in trilobite evolution occurred in the Devonian of the southern hemisphere. Back about 390 million years ago, the southern continents (South America, Africa, Australia, Antarctica—and even India, which later drifted north of the equator and slammed into Asia) were all connected into the supercontinent of Gondwana. There, in the frigid shallow seas that flooded much of these continents, a veritable explosion of trilobites, all from a single root stock, took place in the relatively brief time of 10 or 15 million years. Some very bizarre, highly specialized species appeared during this interval: testimony to high rates of speciation and the rapid accumulation of evolutionary change. [Plate 94, 96]

Ecology, the moment-by-moment interplay between organism and environment (including other organisms), is the guiding principle that controls the rates of evolutionary change. By looking at living fossils, we gain some insight into what causes extraordinarily slow rates of evolutionary change. George Gaylord Simpson, the most influential paleontologist of the mid-twentieth century, focused on these slow rates to help him understand how extremely rapid evolution can occur—particularly instances where change is so rapid that there is little trace of it preserved in the fossil record. The principle can be extended even further when we come to realize how crucial the ecological behaviors of species are in determining how rapidly their adaptations are likely to change in future evolution—and how likely such species are to become extinct and to give rise to descendant species. [Plate 95]

There is much about evolution to be learned from ecological behaviors, as organisms put their evolutionary adaptations to use in making a living. But organisms are only part of the story of evolution. Evolution is the consequence of the interaction between organisms and their environment. In turning now to mass extinctions, our attention becomes more directly focused on the environmental half of the organism-environment equation. The effects of extinction on evolution are only recently coming to be understood. Because extinction has been so pervasive and has played such an important role in the history of life, we should pause once more in awe of those "living fossils," members of ancient lineages that have not only managed to survive through hundreds of millions of years, but have done so using much the same set of adaptive "tools" at their disposal since time immemorial.

CHAPTER 6

EXTINCTION

PLATE 97

PTEROSAUR RHAMPHORHYNCHUS*A ATTACKED BY A RAYFINNED FISH *ASPIDORHYNCHUS.

Upper Jurassic (*ca.* 155 million years). Messel Shales, Eichstatt, Bavaria

The fossil record reveals a great array of animals and plants now completely vanished from the living world.

Credit: Helmut Tischlinger, Jura-Museum Eichstätt.

The cataclysm was as sudden as it was devastating. It came literally out of the blue, shattering the tranquility of Earth's ecosystems. One moment, it was business as usual in the dinosaur communities. The next moment, it was total chaos. A massive comet careened into Earth sending tons of dust and gases up into the atmosphere. Forest fires scorched Earth, and the sun was virtually blacked out. No part of Earth escaped: the creatures that survived the initial blast found their habitats severely disturbed. The dark clouds blanketing Earth lasted for years, blocking so much sunlight that plants could barely photo-synthesize anymore. Whole forests died off. And the tiny photosynthesizing algae floating around on the ocean surface likewise succumbed in great numbers. With the cornerstone base of the food chain effectively put out of commission, the remaining animals—the plant eaters, as well as their carnivorous predators, followed close behind. And then, as the dust was literally settling, leaving the surviving species to regain some equilibrium and set up new ecosystems, disaster struck again. And again. Earth had run into a storm of extra-terrestrial bodies—and life was teetering on the brink of complete annihilation. [Plate 97, see page 153.]

Science fiction? Sounds like it, but this scenario of events ending the Period, 66 million years ago, comes not from the lurid imagination of a fantasy writer, but rather from a Nobel prize-winning physicist and his team of highly competent—and very serious—scientists. Their picture of the cataclysmic demise of the dinosaurs, fueled by the insatiable curiosity that has surrounded dinosaurs since their nineteenth-century discovery, brought the subject of extinction to the forefront: the 1980s became the "extinction" decade. The public was enthralled, especially by reports of dramatically awful catastrophes doing in the dinosaurs. [Plate 98] The scientific community awoke from a long period of dormancy to confront the phenomenon of mass extinction as a high priority item. And it was no mere coincidence that the intense examination of mass extinctions of the remote geological past coincides exactly with a heightened awareness that species are becoming extinct on a daily basis right now. We may well be in the midst of a mass extinction, one in which we ourselves—our species *Homo sapiens*—seems to be playing the unenviable role of arch, if not sole, villain. Suddenly, we have all come to realize that extinction is a very important phenomenon indeed.

PLATE 98

THESCELOSAURUS WARRENI, JURASSICA,
AN HERBIVOROUS DINOSAUR, CAST OF FOOT.

Upper Cretaceous (*ca.* 85 million years). Alberta, Canada.

Dinosaurs utterly symbolize mass extinctions of the geological past.

PLATE 98

PLATE 99

PLATE 99

CLUTCH OF DINOSAUR EGGS POSSIBLY OF *PROTOCERATOPS.*

Cretaceous (*ca.* 90 million years).
Flaming Cliffs, Mongolia.

Though one theory of dinosaur extinction holds that their eggs were weakened through a form of geochemical pollution, it is clear that major extinctions events always involve more general ecological factors and affect many different groups of animals and plants at the same time.

Extinction is not news to a paleontologist. It is a fact of life—albeit one that paleontologists up until recently have usually ignored in favor of more "positive" tidings. I myself used to have the feeling that extinction was the "downside," and preferred to focus on the "upside": evolution. But it has long since become clear to virtually every paleontologist that evolution and extinction go hand in hand. The one, in a sense, "needs" the other. It's easy to see how extinction can't occur unless evolution had already produced organisms that could suffer that ultimate fate. But only in the past few years have we come to realize how true the converse is as well: it now is crystal clear that life virtually cannot evolve to any significant further degree *unless* extinction has come along to eliminate a goodly percentage of Earth's living occupants. Thus my own ruling passion—the patterns of evolution—has become deeply entwined with extinction—as we shall see in the next chapter. One can't understand evolution, really, without understanding extinction. And you can't understand extinction without first grasping ecology: the rules that govern how organisms of different species live in the same ecosystems, and the controls that determine how many organisms, and how many different species, can live in any particular habitat area. [Plate 99] The subject is fascinatingly complex. Yet, as we shall see, certain simple generalities do surface. Extinction turns out to be, if not quite as simple as a straightforward cometary collision wreaking earthly havoc would have it, at least reasonably intelligible. It is, fundamentally, a story of ecological collapse. If we know what holds ecosystems together, we can make some pretty shrewd estimates of what triggers their collapse. [Plate 100] But this gets the cart before the horse: we need to know more about actual extinctions of the geological past, so we can tackle theories of what causes extinction in the first place.

TIME LINES IN ROCK

We have already encountered the geological time scale. It is an indispensable, if formidable, system of names of divisions of geologic time. It allows us to bracket fossils in time, and was especially important in the days before precise radiometric dates were available. But even now, we can radiometrically date fossil-bearing sedimentary rocks only in exceptional circumstances. When I tell somebody that a "Middle Devonian" trilobite is approximately 380 million years old, I mean that rocks called "Middle Devonian" have been dated through analysis of radioactive decay in at least one locality in the world. I know the trilobite is "Middle Devonian" in age because it comes from rocks determined to be about the same age as these precisely datable rocks elsewhere. And we know that because fossils occur in a regular, stratified order through time: rocks are about the same age if they contain the same, or closely similar fossils.

Well and good. Geologists have been determining the relative age of geological strata by looking at their fossil content for almost 200 years now. What we have been somewhat slower to realize is that the divisions of geologic time (worked out, in larger part, by the mid-nineteenth century) are based on actual "packages" of living systems—intervals of geological time recorded in the rock record by the history of discrete faunas and floras. Many species will appear near the beginning and disappear by the end of such packaged intervals of time—though species may appear or disappear at any time within the interval, and some species, of course, will survive into the next succeeding interval. The best evidence that extinctions have occurred, over and over again, is the chart (See page 13.) showing the subdivisions of geological time over the past 4.5 billion years. Life, in essence, *defines* geologic time. The last 540 million years is divided into the familiar Paleozoic, Mesozoic and Cenozoic Eras: ancient, median and recent life. These, the three grand packages of complex life, are demarcated by two of the greatest mass extinction events yet to have hit Earth. [Plate 101]

PLATE 100

SMILODON CALIFORNICUS,
A TRUE SABERTOOTHED CAT

Pleistocene (*ca.* 25,000 years). Rancho La Brea, Los Angeles, California.

The large mammal fauna of North America as recently as 10,000 years ago was as rich and varied as that of the African plains today.

PLATE 101

KAPROSUCHUS SAHARICUS,
A SAHARAN CROCODILE.

Upper Cretaceous (*ca.* 93 million years). Gadoufaoua, Agadez District, Niger Republic.

Though obviously crocodilian, this species (the name means Saharan "boar crocodile") belongs to a lineage of crocodilians that did not survive the end-Cretaceous extinction.

Credit: Paul Sereno, University of Chicago.

The event that closed the Paleozoic was the worst. Paleontologist David M. Raup, of the University of Chicago, has calculated that perhaps as many as 96% of *all species on Earth at that time* may have become extinct! The end-Cretaceous event, more famous because that is when the dinosaurs finally disappeared, was nonetheless far less devastating. But changes in the complexion of the world's ecosystems—particularly the dinosaur-dominated terrestrial ecosystems—was nonetheless so marked that early geologists had no trouble at all drawing a line between "middle" and "modern" ages of life right at the point when the dinosaurs on land, and the ammonites (shelled relatives of squid and octopi) of the seas, disappeared. Because mammals soon replaced dinosaurs as the large-bodied vertebrate constituents of terrestrial ecosystems, the basic structure of ecosystems as they are still familiar to us today was established, and the modern age of life had begun—some 66 million years ago.

But the geological time scale has many more divisions than the simple tri-fold Paleozoic, Mesozoic and Cenozoic scheme would suggest. Some are world-wide in extent—such as the internationally agreed-upon major divisions of the three Eras into twelve "Periods." In turn, each of the periods is further subdivided into Epochs—though here the divisions begin to differ depending upon where on Earth you are looking at them. All this means, in a nutshell, is that true and often global mass extinctions are very real phenomena indeed. And it also means that some extinctions have been more massive, more pervasive in extent than others: they have covered the globe, and affected virtually every major lineage, and every basic kind of ecosystem, in existence. Other events are more minor, striking only some regions, or some habitats, or even just some groups, rather than others. Not surprisingly, these less severe events are more common—and serve as the basis of locally recognized subdivisions of geological time. [Plates 102, 103]

PLATE 102

PLATE 102

TIMORECHINUS MIRABILIS,
AN UNUSUAL SEA LILY (CRINOID ECHINODERM).

Upper Permian (*ca.* 250 million years). Island of Timor, Indonesia.

The mass extinction at the end of the Permian Period was the most devastating of all, killing off these strange crinoids and many other species.

PLATE 103

DIADECTES SP.,
AN AMPHIBIAN.

Lower Permian (*ca.* 270 million years). Wichita Basin, Texas.

The Permo-Triassic revolution affected life in all ecological realms.

PLATE 104

***MACHAEROPROSOPUS* SP., A SUPERFICIALLY CROCODILE-LIKE REPTILE.**

Upper Triassic (*ca.* 215 million years). Arizona.

Extinction claimed many groups on both land and in the sea at the end of Triassic times. Phytosaurs were among those that succumbed.

Six major global extinction events stand out over the past 540 million years. In order of occurrence, they are (1) Upper Cambrian (roughly 485 million years ago), (2) Upper Ordovician (*ca.* 445 million years ago), (3) Upper Devonian (some 370 million years ago), (4) Upper Permian (end Paleozoic—252 million years ago), (5) Upper Triassic (201 million) and (6) Upper Cretaceous (end Mesozoic—around 66 million years ago). Others are noteworthy as well. Especially important for us are the Late Pleistocene (Ice Age) extinctions, predominantly of large mammals—an event that may well still be underway. An event, moreover, that is at least partly attributable to the hunting and habitat-destroying activities of *Homo sapiens*—our very own species. The degree to which we have contributed to these extinctions is crucial to our understanding what effects we are having on the world's ecosystems around us right now. The study of extinctions of the remote geological past sheds light on how extinctions occur without our help. Ice Age extinction is the transitional case, where humans enter for the first time into the extinction picture. The urgency for understanding all these events is clearly crucial to evaluating what is going on in the here and now. [Plate 104]

We need to know if there is any common thread to all these extinctions. Let's focus, first, on events at the end of the Cretaceous—not just because the demise of the dinosaurs has made it is the most celebrated case, but because the intensity of work in the last decade makes it the best documented and understood episode of all the mass extinction events. And let us ask, first, what made Luis Alvarez (the Nobelist in physics) and colleagues devise that unearthly scenario of cometary impact with which we started this chapter.

Walter Alvarez, Luis's son, is a geologist. Determined to examine the physical events right before, during and after the boundary between the Cretaceous and Tertiary Periods ("K-T" boundary, for short), he chose an exposure near the Italian town of Gubbio. The rocks there were reputed to hold a particularly complete record of geological time just as the Cretaceous was giving way to the Tertiary. [Plate 105] The boundary itself was fixed by the occurrence of fossils, particularly minute, single-celled shelled amoebae known as foraminiferans. Right at the boundary line, Alvarez collected samples of a .5 inch thick red clay layer—innocuous-looking enough, but nonetheless something of an anomaly, stuck in as it was amidst a thick sequence of limestones.

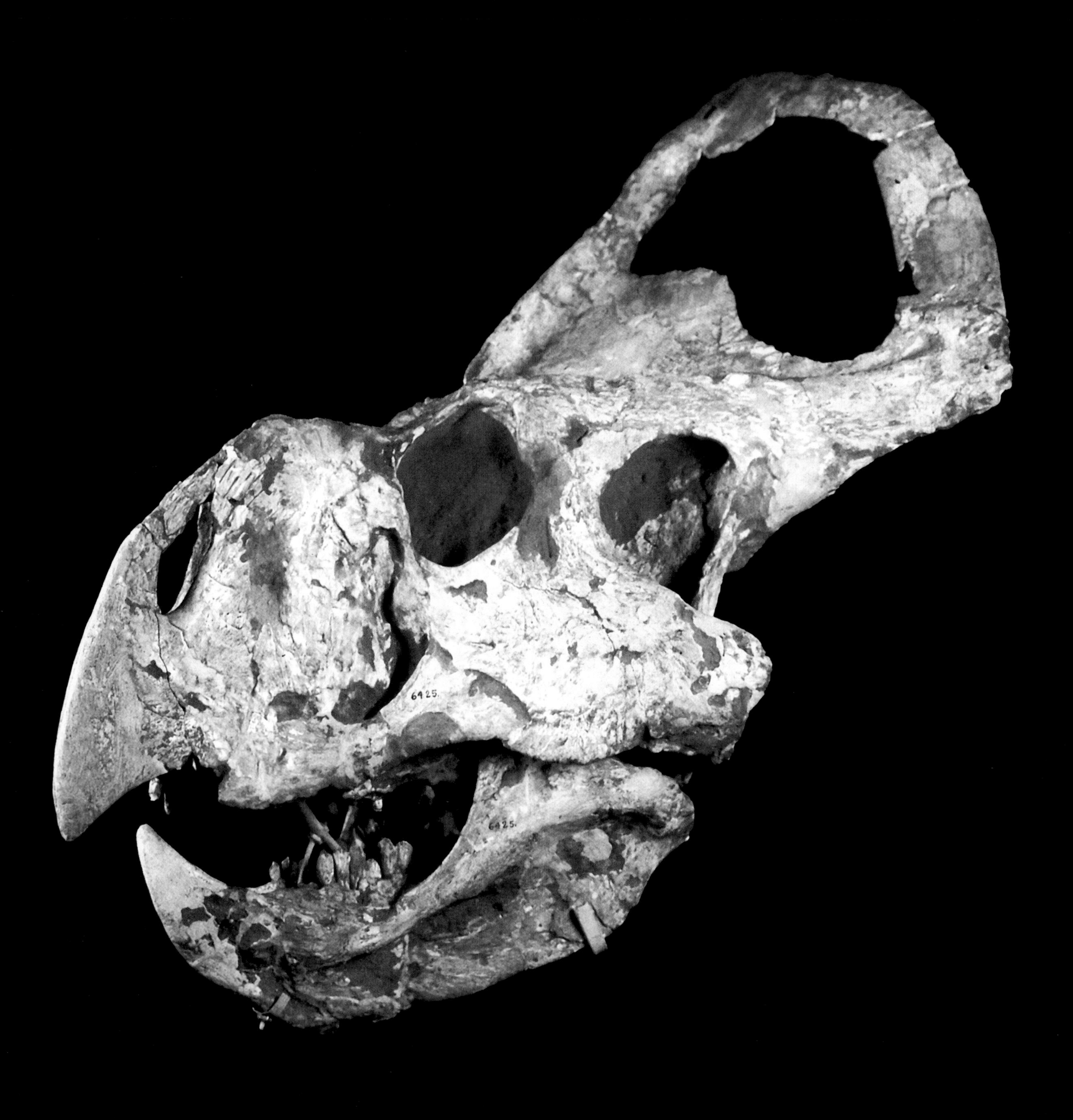

PLATE 105

Back in the lab, the Alvarez team was astounded to find that iridium, a rare metallic element akin to platinum, occurred in the red clay in concentrations far higher than is normally expected for rocks near Earth's surface. This obscure element, iridium, came across most clearly as a complete anomaly. As the senior Alvarez proclaimed, only meteorites have ever shown such high densities of iridium. We now know that iridium occurs in dense concentrations deep within the bowels of Earth—and that volcanoes may belch up similar high concentrations from time to time.

But, in the early 1980s, it seemed a safe bet that the iridium virtually *had* to have come from outer space. And thus was born the "Alvarez hypothesis": something from outer space, a comet or meteor (a "bolide") impacted somewhere on Earth's surface, sending up that tremendous cloud of dust and steam. The initial shock would have killed many organisms wherever that bolide landed, and the search continues for an ancient crater that might correspond to the impact. Because over 70% of Earth's surface (more, actually, in the Cretaceous) is covered by water, and because craters would not have formed in the truly deep sea abyssal plains, the crater may never be found—there may well be no actual crater to find. Nonetheless, candidate craters are from time to time proposed. A particularly likely suspect is a crater just offshore from the Canadian Maritimes, said to have been formed at just about the right time, towards the end of the Cretaceous. Another is Chixulub crater in the Gulf of Mexico, just off the Yucatan Peninsula.

There are other ways to test the "impact hypothesis." One would expect to find iridium "anomalies" at other sites where the K-T boundary lies exposed to view—or can be penetrated, for that matter, by subterranean drilling. And find them one does: all around the Mediterranean Basin, up in Denmark, and even in places as far afield from Italy as western North America where, tellingly, iridium shows up at the K-T boundary in what were terrestrial (as opposed, that is, to marine) environments.

Most meteors burn up in the atmosphere before ever striking Earth. When large objects do manage to get through with enough matter left to crash into the surface, they produce telltale mementos other than craters. Crystals of "shocked quartz," with fracture lines that can only be produced by high velocity impact, are one such vestige of past collisions. Shocked quartz is now being detected in the boundary clays and other sediments of the K-T boundary in a number of locales. Because high iridium levels *can*, at least potentially, be produced by intense volcanic activity; and, because there is growing evidence that there was in fact *very* intense volcanic activity towards the end of the Cretaceous, some geologists actively contend that it was indeed volcanoes, *not* meteors or comets, that produced the iridium anomalies seen in so many sites, most notably the Deccan Traps of west-central India. But the shocked quartz, it seems, tips the evidential scales towards the original Alvarez notion. And there are other signs, as well, that the impact hypothesis is valid. For example, floral evidence seems to bear out the hypothesis of raging wildfires. The fossil record of pollen grains at many terrestrial sites shows ferns vastly outnumbering flowering plants just above the K-T boundary. Ferns are known to be the first plants to come back after wildfires have destroyed living vegetation. In sum, the evidence is pretty convincing that some large object indeed must have collided with Earth near the end of the Cretaceous.

Were that all to the impact hypothesis, we could turn straight away to the actual patterns of extinction of Cretaceous species to see how they fit the impact scenario. But, in the manner of true scientific enquiry, particularly when a topic is "hot" and attracts a lot of attention, the impact scenario quickly grew as a general theory of extinction. Not just the end-Cretaceous mass extinction, in other words, but conceivably all mass extinctions (certainly all global ones) might just be the outcome of extraterrestrial incursions. One major boost to this notion came from paleontologists Jack Sepkoski and the aforementioned David Raup, who, analyzing Sepkoski's vast computerized data base of the fossil record, concluded that extinctions are not randomly occurring events, but in fact seem to come with a clock-like regularity, roughly every 26 million years.

Such epic regularity virtually demands astronomical causes. Only in the heavens, specifically in the orbits of planetary and solar bodies, can we hope to find a source of periodicity on the order of 26 million years. Astronomers sprang to work immediately and offered various suggestions on possible heavenly mechanisms. Most meteors come from the "asteroid belt" between Mars and Jupiter. Like the nine planets and their satellites, asteroids occupy regular orbits around the sun. Comets, on the other hand, have eccentric and wildly elliptical orbits. The "Oort cloud" is a hypothetical swarm of cometary bodies in the very farthest reaches of the solar system. Every so often, a comet comes hurtling through the inner reaches of the solar system, swings closely around the sun, and then hurtles back into those outer reaches. Astronomers quickly concluded that a mechanism that

PLATE 105 (See page 165.)

***PROTOCERATOPS ANDREWSI*, AN HERBIVOROUS HORNED ("CERATOPSIAN") DINOSAUR.**

Cretaceous (*ca.* 90 million years). Flaming Cliffs, Mongolia.

This species is a primitive relative of the great *Triceratops,* which lived at the very end of Cretaceous times, along with the even more famous *Tyrannosaurus rex.*

would hurl objects earthwards at regular intervals would be easier to imagine for comets than for asteroids. Thus comets displaced meteors as the extraterrestrial visitor of choice in the impact hypothesis.

What could perturb the Oort cloud on a regular basis? Astronomers quickly posited that some as yet undetected celestial body could conceivably dislodge comets as its orbit periodically brought it close to the Oort cloud. Two candidates quickly emerged. "Planet X," as it was pun-ishly dubbed, is a predicted tenth planet. It has yet to be found. "Nemesis," on the other hand, is a predicted sister star to our own sun. Many stars occur as twins, revolving around some central point. Optical astronomers call them "doubles." Our sun's twin would be a remote, dim star not expected to have been noticed before someone actually went looking for it. They've been looking for it for some time now. So far it has avoided detection.

Meanwhile skeptics voice other doubts. Some challenge the very extraterrestrial origin of the iridium layer—though, as we have seen, so far the evidence does seem to support a bolide impact at the end of the Cretaceous. Actually, some exposures show more than a single layer, and the suspicion is growing that more than one impact may have occurred—as if Earth had encountered a "swarm" of extraterrestrial objects. A more serious problem (beyond that is the failure to locate a physical cause for the supposed periodic waves of comets careening through the solar system) is the doubt that a number of paleontologists have that mass extinctions in fact do occur with true periodicity. The debate centers on the nature of the data base—and the statistical techniques used in the analysis. In fairness, it can only be said at this juncture that the case for periodic incursion of lethal objects from outer space is not proven.

The plot thickens when we look at the boundaries marking the other five most severe extinction events since the advent of complex life. Geologists have detected a high iridium content at the Permo-Triassic boundary (*i.e.* the Paleozoic-Mesozoic boundary event—the biggest extinction event so far). Similarly, high iridium concentrations have been found at the Triassic/Jurassic boundary. So far, that's it. [Plate 106] The evidence so far seems to show that, if there is to be a single, all-encompassing theory of mass extinction, it will not prove, after all, to involve rocky visitors from outer space. We must take another look at Earth itself, specifically to the forces on Earth that shape ecosystems, for the seeds of a truly all-encompassing theory of mass extinction. To get at that problem, let's start with a closer look at what actually happened before, at and right after the K-T boundary "event."

PLATE 106

***SALIX SP.*, LEAF OF A WILLOW TREE.**

Upper Eocene or Lower Oligocene (*ca.* 35 million years). Near Clarno, Oregon.

Extinctions of a less drastic scale than the six greatest global events punctuate the history of life on Earth throughout geological time—including a series of events around the Eocene-Oligocene boundary.

PLATE 106

PLATE 107

PLATE 107

***PLACENTICERAS INTERCALARE,* AN AMMONITE CEPHALOPOD.**

Upper Cretaceous (*ca*. 80 million years). Near Medicine Hat, Alberta, Canada.

The ammonoids survived the major mass extinctions in the Devonian, Permian and Triassic—only to succumb, at last, to the ecological collapse at the end of the Cretaceous.

PATTERNS OF MASS EXTINCTION

Some paleontologists resisted the Alvarez "bandwagon" ("steamroller" might actually be more apt) as extinction grew into a major scientific hot topic in the early 1980s. Not that anyone doubted the existence of the iridium anomaly. Out of sheer curiosity, I had it checked myself, after collecting some samples of the green "Fish Clay" (equivalent to the "red clay layer" at Gubbio) at Stevn's Klint in Denmark. Sure enough, the iridium (and platinum to boot) was virtually "off the scale." It wasn't the iridium—it was the actual pattern of extinction that bothered paleontologists].

It had been known for years that there had been a major faunal turnover as the Cretaceous gave way to the Tertiary. [Plate 107] What worried many paleontologists (myself certainly included) was that it had also been known that the extinction event had been well underway long before the actual K-T boundary layers had ever been deposited. Experts in various different groups of Cretaceous fossils immediately started to object: the Alvarez scenario, at least in its earliest guise, saw the K-T extinction as an almost literally overnight occurrence. The process would have taken a few years, and, at most, decades, or perhaps even a century or two. But not millennia—still less the 500,000 years that Cretaceous paleontologists were invoking as the minimum amount of time that the data seemed to suggest the extinction had taken. Some called for even more time: dinosaurs, arch symbols of the entire extinction event, had been on the decline for much of the Upper Cretaceous—for upwards, that is, of 3 million years. No way, many paleontologists thought, can we cram all these events into the instantaneous framework proposed by the Alvarez hypothesis.

What is the resolution? Paleontologists have done a lot of additional detailed field studies since the Alvarez hypothesis was put forward. One very revealing study, by paleontologist Greta Keller, was on the foraminiferans (shelled amoebae) at the Tunisian site of El Kef. The results seem to confirm a hybrid combination of traditional paleontological viewpoints with elements of the Alvarez scenario. Specifically, eight species of foraminiferans disappear right at the boundary—which is marked, not just by their disappearance, but by an iridium anomaly as well. But Keller showed that there had not been one, but at least *four* such events: some 25 centimeters below the iridium layer, there had been an earlier extinction event that saw six species disappear. And there were two additional extinction waves *after* the iridium layer at the K-T boundary: twenty-two species drop out at two different horizons within 7 centimeters *above* the iridium layer! Though the data fit a revised version of the Alvarez hypothesis, calling for a series of "multiple impacts," they also take away the idea that mass extinctions are single, overnight incidents.

And there is no doubt that dinosaurs and other groups—notably the ammonoids (as symbolic of extinction in the seas as dinosaurs are of extinction on land) had been undergoing a progressive decline that expands the duration of extinctions even more than the data from El Kef seem to suggest. It is a mistake, of course, to worry only about the extinction of dinosaurs, or the ammonoids, or any other particular group, when we know that extinction cuts across taxonomic boundaries, taking out many unrelated groups at the same time. Extinction is an ecological affair. We seek to explain, not the disappearance of this group or that, but the decline of regional ecosystems, which severely affected and even drove to complete extinction, many different plants and animals of diverse identity and ecological habit. Nonetheless, the data of individual groups are important to establishing the actual total pattern of extinction. And, as we have already seen, dinosaurs had been declining markedly and steadily for at least 3 million years prior to their actual extinction.

Among the ammonoids, several longstanding groups were showing a similar decline from the Middle Cretaceous on. Only the scaphites, a young group of rather oddly shaped ammonoids, was showing signs of active evolution. They appear to have actually been increasing in diversity as the Upper Cretaceous wore on to a close. Ironically, scaphites figure prominently in early theories of extinction. Because scaphites depart from the normal logarithmic coil of most ammonites, they were taken as a sign that ammonoids had run out of evolutionary steam and had fallen prey to an evolutionary form of old age—so-called "racial senescence." [Plate 108] Skipping up to Ice Age extinctions of the past 10,000 years for a moment, the "Irish Elk" (a species of giant deer, actually) was long thought to have evolved antlers so huge that they interfered with the ability of males actually to go about the normal process of living—and so contributed directly to the demise of the species. The Irish Elk story has often been cited as another example of "racial senescence." All such examples have been debunked: the arguments that the individuals bearing such adaptations were actually dysfunctional has been convincingly refuted. And, more relevant to our theme, it is by now abundantly clear that scaphitid ammonoids and the Irish elk became extinct along with an impressive and truly diverse cadre of other species living in all sorts of different ecological settings.

PLATE 108

***DIDYMOCERAS* SP., AN ABERRANTLYCOILED AMMONITE CEPHALOPOD**

Upper Cretaceous (*ca.* 80 million years). Black Hills, South Dakota.

Because of its deviant coiling pattern, ammonites like Didymoceras, and the scaphites were once taken as evidence of "racial senescence," or extinction caused by decline in the genetic integrity of a lineage.

PLATE 108

Some of the Cretaceous extinctions did indeed embrace highly specialized, even odd, organisms. Rudistid clams, briefly encountered in Chapter 2 as mollusks that evolved convergently on corals, were major reef-builders in later stages of the Mesozoic. They failed to get through into the Cenozoic. But most of the victims were just yeoman species—numbers of clams, snails, grasses, dinosaurs, flying and marine reptiles, and so forth. No group was left completely untouched. And some saw their entire remaining "standing crop" of species snuffed out—thus higher groups, such as ammonoids and dinosaurs, failed to get across the K-T boundary.

Yet, clearly, extinction (thankfully!) failed to claim *every* species. One outstanding puzzle focuses on the survival of some species, while close relatives (or physiologically and ecologically similar) species fell victim to the extinction scythe. Take dinosaurs. Birds are really nothing more (nor less!) than specialized dinosaurs. Evolving in the Jurassic, birds made it through into the Cenozoic just fine. [Plate 109] As did crocodiles, after birds, the closest relatives to the dinosaurs. Current research is focusing on different habitat preferences of surviving groups in an effort to sharpen the picture on just who does survive mass extinctions—and why.

PLATE 109

PTEROSAUR FOOTPRINT, POSSIBLE TAPEJARID OR TOOTHLESS PTEROSAUR.

Upper Jurassic (*ca.* 150 million years). Villaviciosa, Spain.

Birds survived, while other forms of flying reptiles did not make it across the Mesozoic-Cenozoic boundary.

Credit: José Martinez Carlos García Ramos and Laura Piñuela.

Thus the K-T extinction picture is not as neat and clean as proponents of the impact hypothesis would have it. Some paleontologists in recent years have argued that a few dinosaurs actually did survive into the Paleocene. Something, assuredly, supplied all that iridium, and chances are it was from one (or a series of) bolide impacts. But ecosystems had been in trouble for a good while before the terminal events that seem to have ended the Cretaceous with a bang. We will gain a deeper understanding of what underlies mass extinctions if we look beyond the events that ended the Cretaceous. The patterns developed at the other five of the biggest mass extinctions in geological time point the finger, not at the heavens, but directly at Earth-bound causes of mass extinction.

The end of the Cambrian Period witnessed a major cut-back of the ruling denizens of the sea, the trilobites. Though they would live on for another 265 million years, never again would trilobites be as diverse or so dominant in the marine communities of the shallow seas that perennially washed over the interiors of Earth's continents. The extinction that ended the Cambrian was but the latest of several that had occurred in the Upper Cambrian. What caused the cut-back, though, is obscured in the mists of time. Sea-level change apparently rearranged habitats to a degree that many local ecosystems were severely disrupted, a theme we will encounter again with greater clarity as our survey of the greatest of mass extinctions proceeds to more recent—and more readily interpretable—events.

The major extinction at the end of the Ordovician was, by all accounts, much more severe than the one at the end of the Cambrian. It also tells us more about what the patterns and underlying causes of mass extinctions might generally be. There were two distinct waves of extinction, the first occurring some 1, or perhaps even 2, million years before the second, which came at what is traditionally taken to be the end of the Ordovician. The fauna that typifies the very latest Ordovician marine communities is best known from Europe. Consisting of relatively few species, it represents a fauna adapted to very cold water. The Upper Ordovician South Pole, it turns out, was positioned in modern day North Africa (in what is now the Sahara! The poles appear to have "wandered" in the geological past, not because they themselves were moving around, but because Earth's plates, including its continents, have continually shifted their positions on the Earth's surface throughout geological time.) When poles are located over land masses (as the South Pole is today, smack over Antarctica), extensive and very thick ice sheets are formed. These "continental glaciers" have a truly chilling effect on both oceanic and atmospheric circulation—and global cooling is the usual result. Most paleontologists are agreed that the Ordovician extinctions are directly related to a significant episode of global cooling associated with significant glaciation.

Global cooling is a theme that in fact is repeated in virtually all other mass extinctions that have received close study. It emerges immediately in any consideration of the next great extinction—at the end of the Devonian. The major pulse of extinction there, actually, occurs between the last two divisions of the Devonian, between the "Frasnian" and "Famennian" Epochs, just to be precise. (Geological age names are generally based on geographic names of places where rocks of the relevant age were either first recognized, or are best developed. Frasne and Famenne are towns in the Belgian countryside). Just as in the Ordovician, very latest Devonian faunas consist of rather few species, living in oxygen starved and evidently very chilly waters. The glass sponge *Hydnoceras*, for example, lived in Upper Devonian waters with very few other invertebrate species. [Plate 110] Modern glass sponges typically live in the frigid oceanic abyss.

PLATE 110

HYDNOCERAS TUBEROSUM,
A GLASS SPONGE.

Upper Devonian (*ca.* 365 million years) Olean, New York.

Modern glass sponges betoken cold temperatures, and the presence of this sponge in Upper Devonian environments may reflect a cooling of oceans towards the end of Devonian times.

When we reach the "Permo-Triassic event," the greatest extinction (so far!) of them all, we meet headlong an important debate among paleontologists seeking earthly causes for global events of mass extinction. Recent research indicates that this massive extinction event took place within the incredibly short span of 60,000 years. Yet we seem to be no closer to a definitive understanding of the cause—or causes—of this massive turnover of Earth's biota. Rather, the recent history of the debate sheds important light on general factors that underlie all extinction events—including the accelerating loss of species in the modern world. Though most recent work on the Permo-Triassic extinction event invokes complex sets of causes, earlier studies focusing on global temperature change or loss of habitat area—while now considered erroneous or overly simplistic—nonetheless crystallize the factors that cause extinctions in important ways. [Plate 111] Until very recently, it was generally believed that, if anything, Earth was getting *warmer* towards the end of the Permian. Extensive salt deposits, for example, imply dry and (some have assumed) rather warm conditions; salt forms as extensive bodies of salty water evaporate. And there had been extensive glaciation earlier in the Upper Paleozoic—without any associated pulse of extinction. Yet, as paleontologist Steven Stanley has pointed out, there is actually new evidence that parts of the globe, at least, were under ice as the Paleozoic grew to a close. He believes that the earlier glaciation failed to set off massive extinctions because the world's biota was still recovering from the massive extinctions that ended the Devonian. In any case, there is at least some evidence linking the late Paleozoic extinction with global cooling.Opponents of the idea that global cooling underlies mass extinction generally prefer the alternate hypothesis that extinction stems directly from reduction, or even outright loss, of habitat space. [Plate 112] As an extreme example, consider what happens when entire inland seas dry up: the habitat simply disappears. If species have nowhere to go, extinction is inevitable—as we saw happened to species of the trilobite genus *Phacops* once their habitats had been eliminated. Paleontologist Norman Newell, who, virtually alone among paleontologists throughout the middle years of the twentieth century, took the phenomenon of mass extinction seriously, has long thought that habitat loss is the key to understanding mass extinction. Whole faunas might easily be severely affected by prolonged habitat disturbance and modification.

Paleobiologist Thomas Schopf, in fact, applied the principles of "island biogeography" in an attempt to show that the Permo-Triassic extinction was the direct result of loss of habitat space. "Island biogeography," a theory developed by ecologist Robert MacArthur and renowned evolutionary biologist and conservationist Edward O. Wilson, maintains that numbers of species present in a given area is a direct and simple reflection of the size of the area. Restricting his analysis to marine invertebrates, whose fossils are known almost exclusively from sediments that floored the shallow water seaways flooding over the continents, Schopf showed that seas had shrunk considerably by the end of Permian times—all one needs, he argued, to accept Newell's thesis that habitat loss is the culprit. Reason enough, say some later paleontologists, to resist the notion that global cooling is the key to mass extinction.

Consider, for a moment, what actually happens when environments change. Take global temperature change: when the glaciers expanded southwards from the north polar region over much of both Eurasia and North America (something that happened at least four times in the past 1.6 million years) the plants that form the very base of the terrestrial food chain perforce had to move. They moved, of course, through seed dispersal. Entire species changed their distributions as a direct reflection of environmental change: cooling, not to mention the physical presence of ice, which destroys everything in its path. Those species whose organisms were able to locate suitable habitat survived—more or less intact, in the same recognizable form, as we saw when we examined the notion of stasis in Chapter 3. Extinction follows such change only when suitable habitats no longer can be found, or "recognized."

Global cooling (or warming, for that matter) acts by *rearranging* habitat space—which very often amounts to the elimination of some habitats as others come to take their place. Paleontologist Elisabeth Vrba has written eloquently of a major climatic event some 2.5 million years ago. She studied the effects a major global cold snap had on the ecosystems of East Africa. She found an abrupt changeover (which she calls a "turnover pulse"), with many species becoming extinct and new ones either migrating in, or actually evolving, to take their place. But the new species take "the place" of the old denizens solely in terms of the physical space they occupy. Ecologically, the old habitats are gone, replaced by new ones: in this instance, extensive woodlands gave way to open savannah grasslands. As a consequence, an entirely new ecological cast of characters took up residence.

In short, it is my conviction that the debate between those who look to pulses of global cooling, and those who see mass extinction

PLATE 111

WHATCHEERIA DELTAE,
AN EARLY TETRAPOD.

Middle Mississippian (*ca.* 340 million years). Delta, Iowa.

Named for a town with the unlikely name of "What Cheer," Iowa (where the man who first discovered these remains was born), this early 3 foot long tetrapod has a mosaic of primitive and advanced features, illustrating the complex nature of the evolutionary process—especially in the early days of an evolving lineage.

Credit: Lance Grande, The Field Museum.

as a direct reflection of habitat loss, are really looking at different aspects of the very same phenomenon. Extinction, without doubt, stems from habitat loss. But perhaps the most profound way that habitats are modified and redistributed, changed and "lost," is through episodes of relatively abrupt and severe global temperature change. Any theory of mass extinction must apply to events on both land and sea. A problem with some earlier studies of the massive Permo-Triassic extinction is that they tended to focus solely on marine invertebrates. [Plates 113, 114] It was always difficult to imagine how shrinkage of shallow water marine habitats would effect habitats on dry land. But how do seas shrink, anyway? Plate motions, it is thought, can affect sea levels. *But sea level can sink a lot faster during a period of glacial growth:* continental ice fields can tie up an awful lot of water in a hurry. Global cooling simultaneously rearranges habitat on both land and sea!

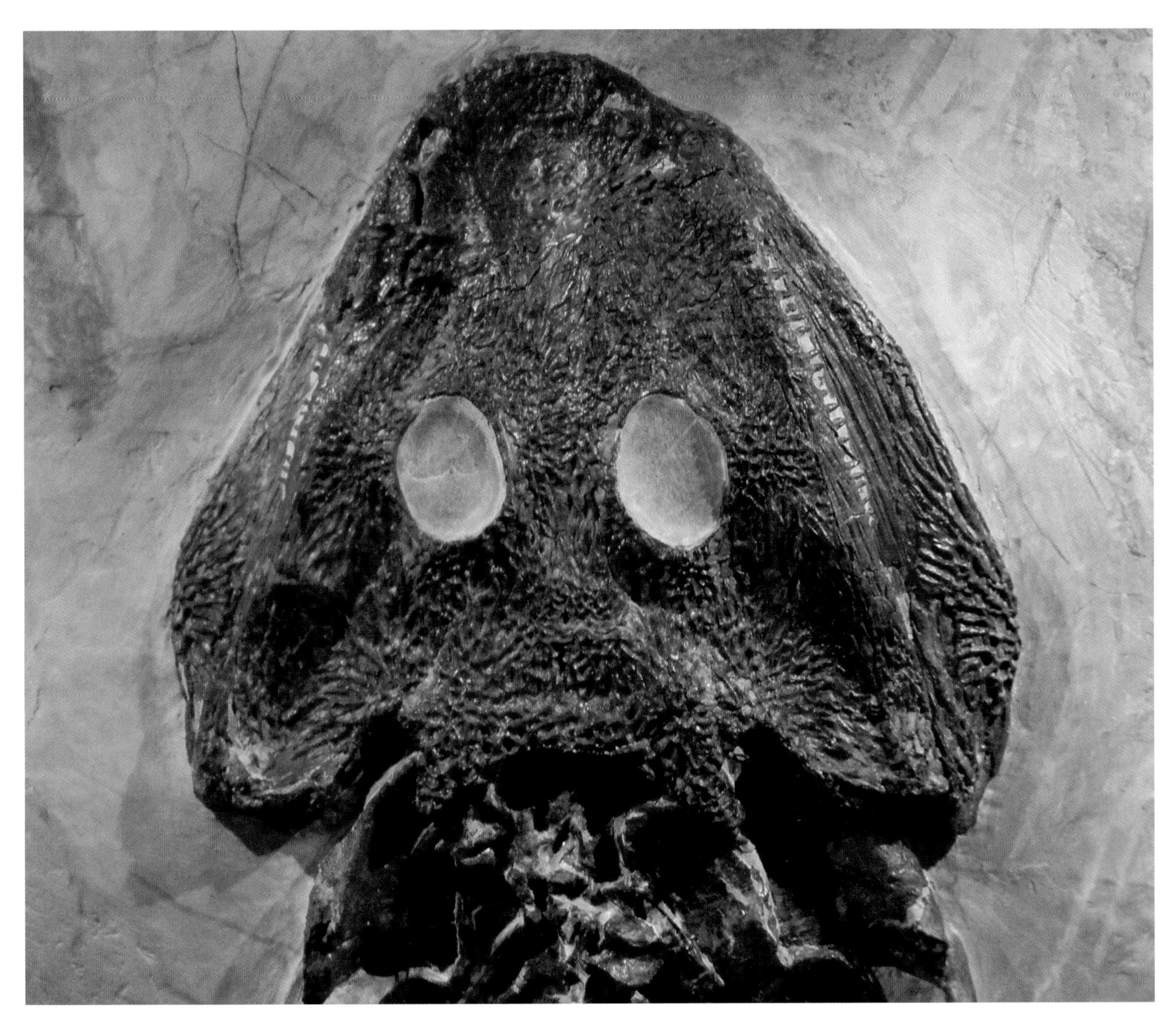

PLATE 112

SCLEROCEPHLEUS HAUSERI,
AN AMPHIBIAN.

Lower Permian (*ca.* 280 million years). SaarNahe Basin, Germany.

Large amphibians like this 3 – 4 foot long *Sclerocephalus* are among the most abundant and (to my eyes) beautiful and arresting terrestrial vertebrates of the Late Paleozoic.

Credit: John Cancalosi.

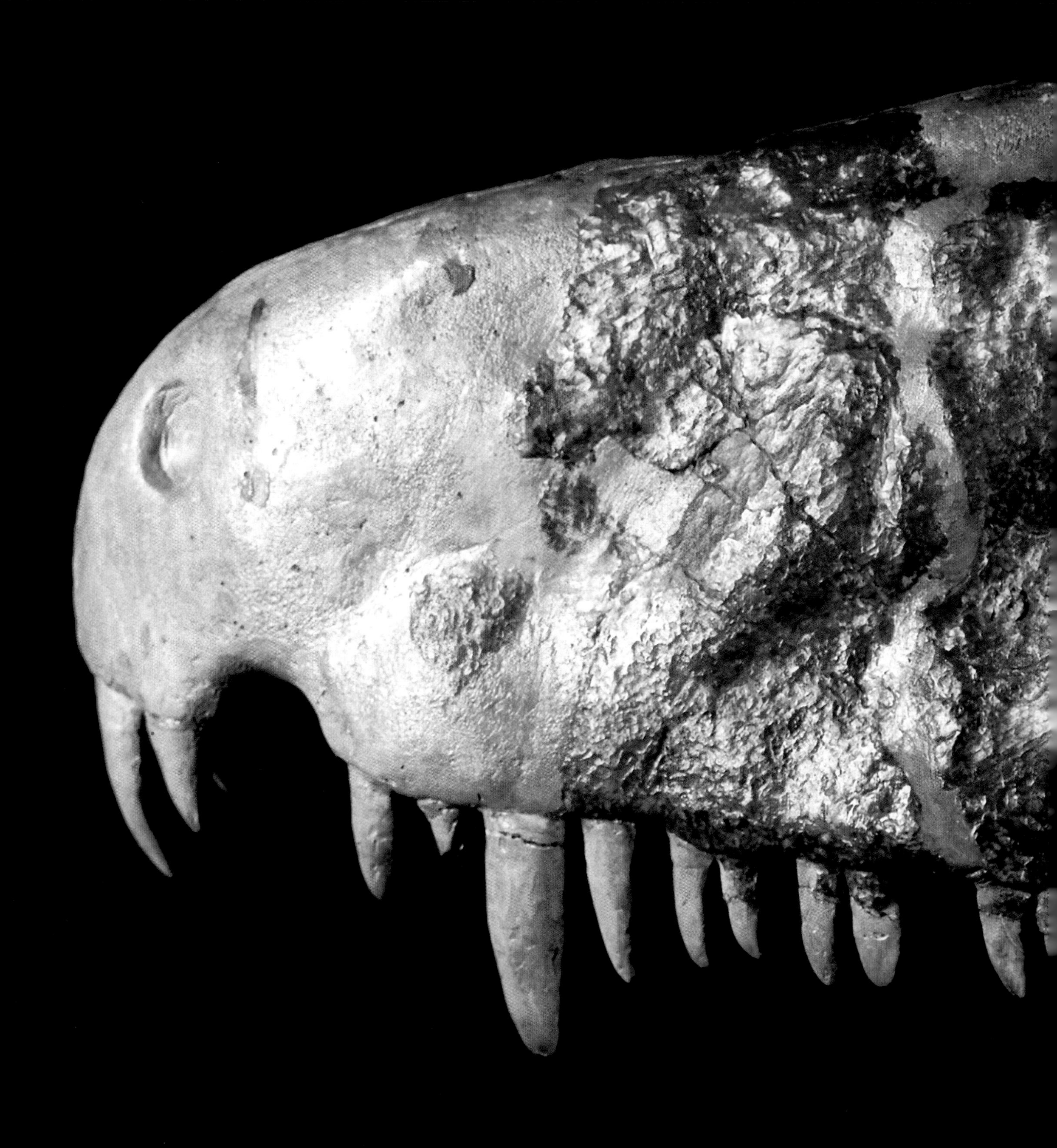

PLATE 113

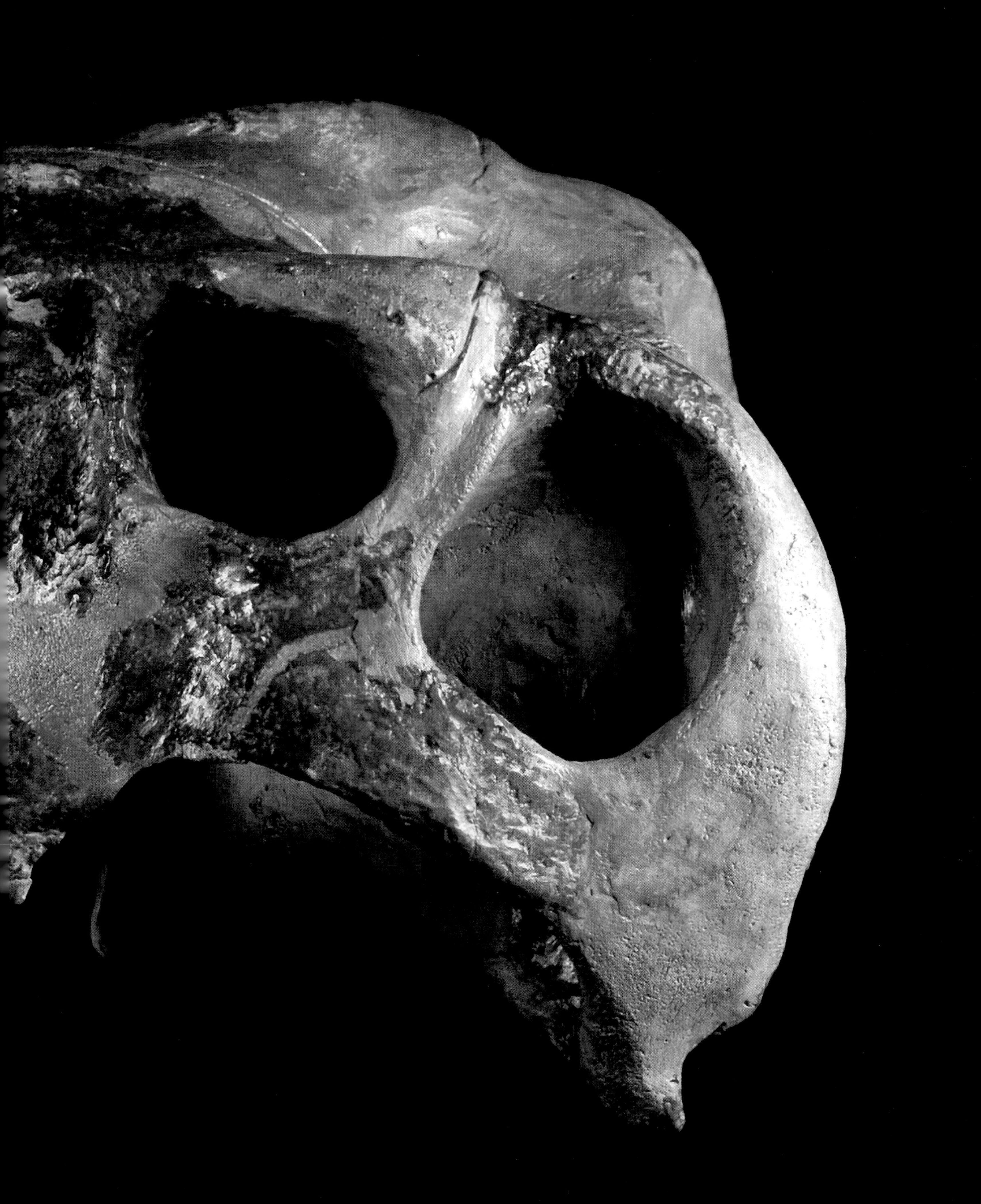

PLATE 114

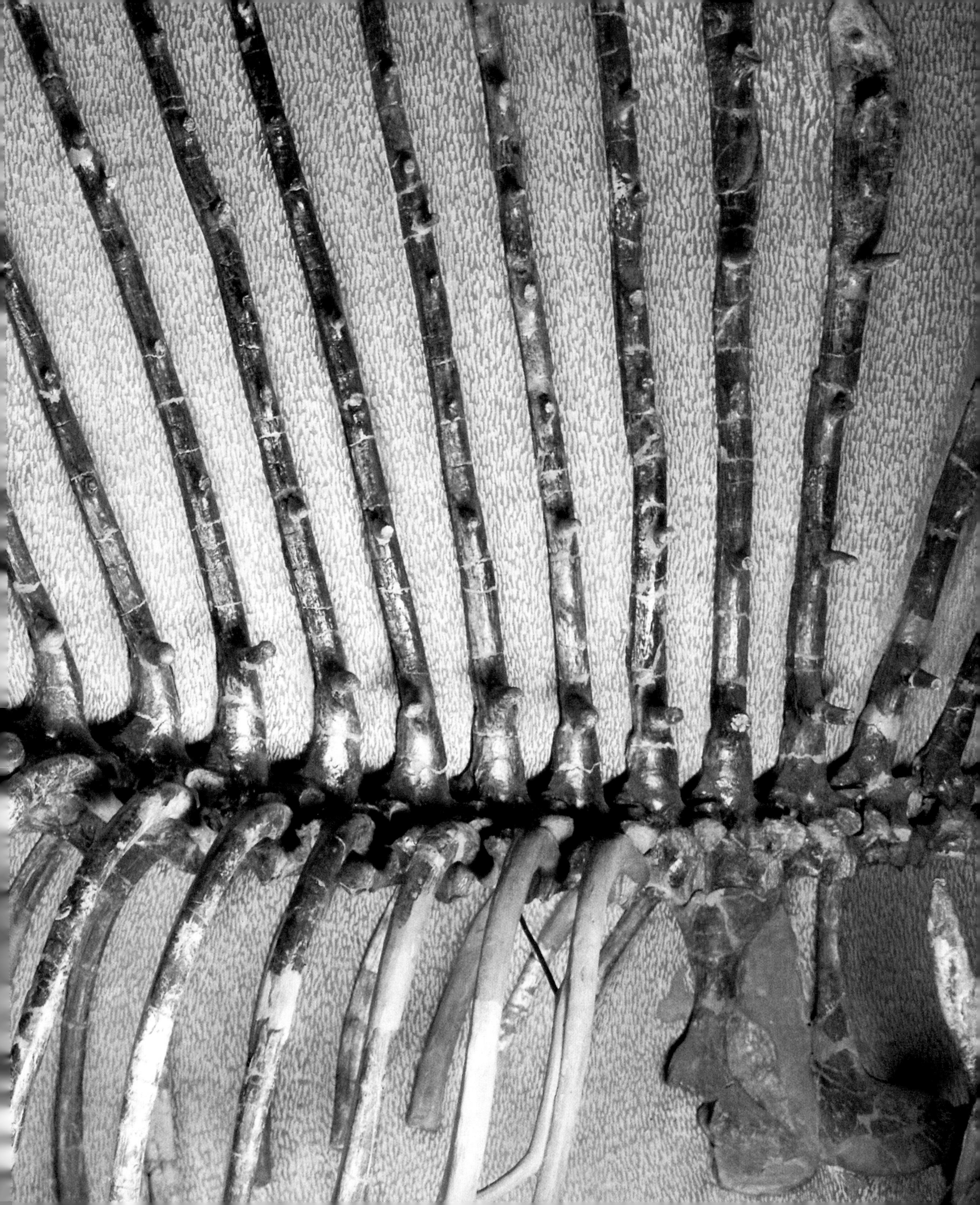

Data for the Upper Triassic (ironically enough, the next succeeding mass extinction) do not so far suggest the occurrence of global cooling. There was extensive habitat change, but its cause so far remains something of a mystery. But how about the K-T boundary? One component of at least some versions of the "bolide impact" hypothesis in fact does call for a sharp drop in global temperatures as a direct effect of the collisions. There is some evidence, moreover, that Earth was in fact undergoing cooling during late Cretaceous times. Though I am no fan of compromise, I really do think that the evidence at this stage compels us to see the mass extinction at the K-T boundary as a hybrid sort of affair: there is little doubt that the world's ecosystems were in trouble for millions of years prior to the events that mark the boundary. It seems quite likely that global cooling lay behind this prolonged decline. But, as we have seen, something produced that iridium layer (in some places, multiple layers), and that *something* seems to have been a massive impact of an extraterrestrial object.

And the effects of such an event (or, as now seems more likely, *series of events*) seems to have been to make a troubled global ecosystem go over the brink. It really seems as if the terminal physical events at the end of the Cretaceous made a bad situation worse—virtually intolerable, in fact.

ICE AGE EXTINCTION—AND OUR CURRENT DILEMMA

As we have seen in greater detail in Chapter 4, our own human evolutionary lineage diverged from the great ape line some 6 to 7 million years ago. For the greater part of those 6 to 7 million years, our ancestral species consisted of rather small bands of hunter-gatherers, each very much a part of their local ecosystems. There is no evidence that early hominin species had any adverse effects on their surrounding environments. Even when *Homo ergaster-Homo erectus* first expanded out of Africa, early humans were still living and hunting in small bands, very much a part of their local environments.

It was when our own species emerged a bit over 150,000 years ago that humans began leaving their mark on the environment. Though it may be hard to believe, large and seemingly exotic mammals were very much a part of the fauna of the late Pleistocene (Ice Age) in both Europe and North America as recently as 8,000 years ago. Woolly rhinos and mammoths, mastodons, saber-toothed cats, giant lion and buffalo—all were common elements of the colder reaches of the northern climes. Many of these species show up in stunning detail in beautiful paintings on cave walls in Spain and France that date back to 16,000 years ago. These large Pleistocene mammals are, of course, no more. [Plate 115] Remnants of this fauna persist in the tropics. African mammals constitute a particularly vivid reminder of these recently bygone faunal days, a sampling of the much more diverse mammalian faunas that covered much of Earth until quite recently.

PLATE 113 **(See page 180–181.)**

DIMETRODON LIMBATIS,
A CARNIVOROUS REPTILE ALLIED WITH MAMMAL-LIKE REPTILES.

Lower Permian (*ca.* 270 million years). Wichita Basin, Texas.

Dimetrodon was a carnivore, part of a complex ecosystem that disappeared entirely as the Permian wore on in what is now the state of Texas.

PLATE 114 **(See page 182–183.)**

EDAPHOSAURUS CRUCIGER,
A HERBIVOROUS REPTILE ALLIED WITH MAMMAL-LIKE REPTILES.

Lower Permian (*ca.* 270 million years). Beaver Creek, Texas.

The spiny backs of both this plant eater, and its close relative *Dimetrodon* (Plate 113) probably served to regulate body temperature but were once thought by some imaginative paleontologists to have been sails enabling these reptiles to traverse the shallow seas of Texas!

PLATE 115

BISON OCCIDENTALIS,
AN EXTINCT BISON.

Pleistocene (*ca.* 25,000 years) Alaska.

Modern North American bison, considerably smaller than this extinct species, themselves very nearly became extinct through overhunting (by settlers, not by Native Americans), escaping only through a farsighted conservation program through the New York Zoological Society, including Theodore Roosevelt.

What happened? Extinction claimed many of the northern mammals as the glaciers were melting—not expanding. There is no way we can blame global *cooling* for the extinction of animals that were so clearly well adapted to surviving the frigid climes of the glacial north. But, for the first time in the story of global extinction, there is another player in the game—another possible cause of extinction. Ourselves.

Anthropologist Paul Martin has long held that it was over-hunting by humans that was responsible, at least in large measure, for the late Pleistocene mammalian extinctions. Martin bases his claim on the coincidence of arrival of humans in the New World at about the time the big mammals started to disappear. In fact, extinction of species is directly correlated with the arrival of humans all over the world. As our population expanded and people began migrating and occupying all available space, there was a direct, and probably inevitable, negative impact on the local fauna. [Plate 116]

Recently, fossil footprints of large Pleistocene mammals (such as *Megatherium* and the giant armadillo-like species known as glyptodonts) have been found at Pehuen Co in Argentina. There are some human footprints also scattered among the many others in these 12,000 year old footprint layers. Younger footprints nearby, dated at around 7,000 years ago, tell a dramatically different story: the footprints are almost entirely of human beings—some with footwear. Only guanacos and rheas (the small camels and "ostriches" still living in Argentina) remain in these younger sediments—strongly suggesting that humans increased in numbers and probably played a major role in eliminating those big lumbering mammals of the South American Pleistocene. [Plate 117]

PLATE 116

MEADOWCROFT ROCKSHELTER, PALEOINDIAN ARCHAEOLOGICAL EXCAVATION.

Pleistocene (*ca.* 19,000 years ago). Cross Creek, Avella, Pennsylvania.

This archaeological site shows the longest sequence of continuous human occupation in the New World. Genetic, linguistic and archaeological evidence suggests that humans migrated from eastern Siberia into the Western Hemisphere in three waves starting perhaps as early as 30,000 years ago.

Credit: J.M. Adovasio, Mercyhurst Archaelogical Institute.

PLATE 117

PARALLEL TRACKWAYS OF THE GIANT GROUND SLOTH *MEGATHERIUM* AND OF AN INDIVIDUAL *HOMO SAPIENS*.

Upper Pleistocene (*ca.* 12,500 years ago). Pehuen Co, Bahia Blanca (in the background), Argentina.

Though occasional records of human occupancy in the New World go back perhaps as far as 30,000 years, humans first arrived in significant numbers in the New World *ca.* 12,500 years ago—roughly the time these amazing trackways of many species of the mammalian "megafauna" were formed. Human footprints are rare at this level; but some 5000 years later, in nearby exposures of more trackways, nearly all the megafaunal species (including the iconic *Megatherium*) are missing—and presumed extinct. Humans, some birds and a mammal species (probably the still existing guanaco) are the only species to have left their footprints at Pehuen Co 7000 years ago. Negative evidence, perhaps, but the pattern strongly suggests that humans had a sudden and deleterious impact of the native fauna when we arrived.

Credit: Teresa Manera, Universidad Nacional del Sur.

PLATE 118 (See page 188–189.)

***PACHYORNIS ELEPHANTOPUS*, A MOA, A GIANT EXTINCT FLIGHTLESS BIRD.**

Pleistocene (over 10,000 years), South Island, New Zealand.

Large, specialized animals are particularly vulnerable when a clever predator and habitat alternator such as *Homo sapiens* appears on the scene.

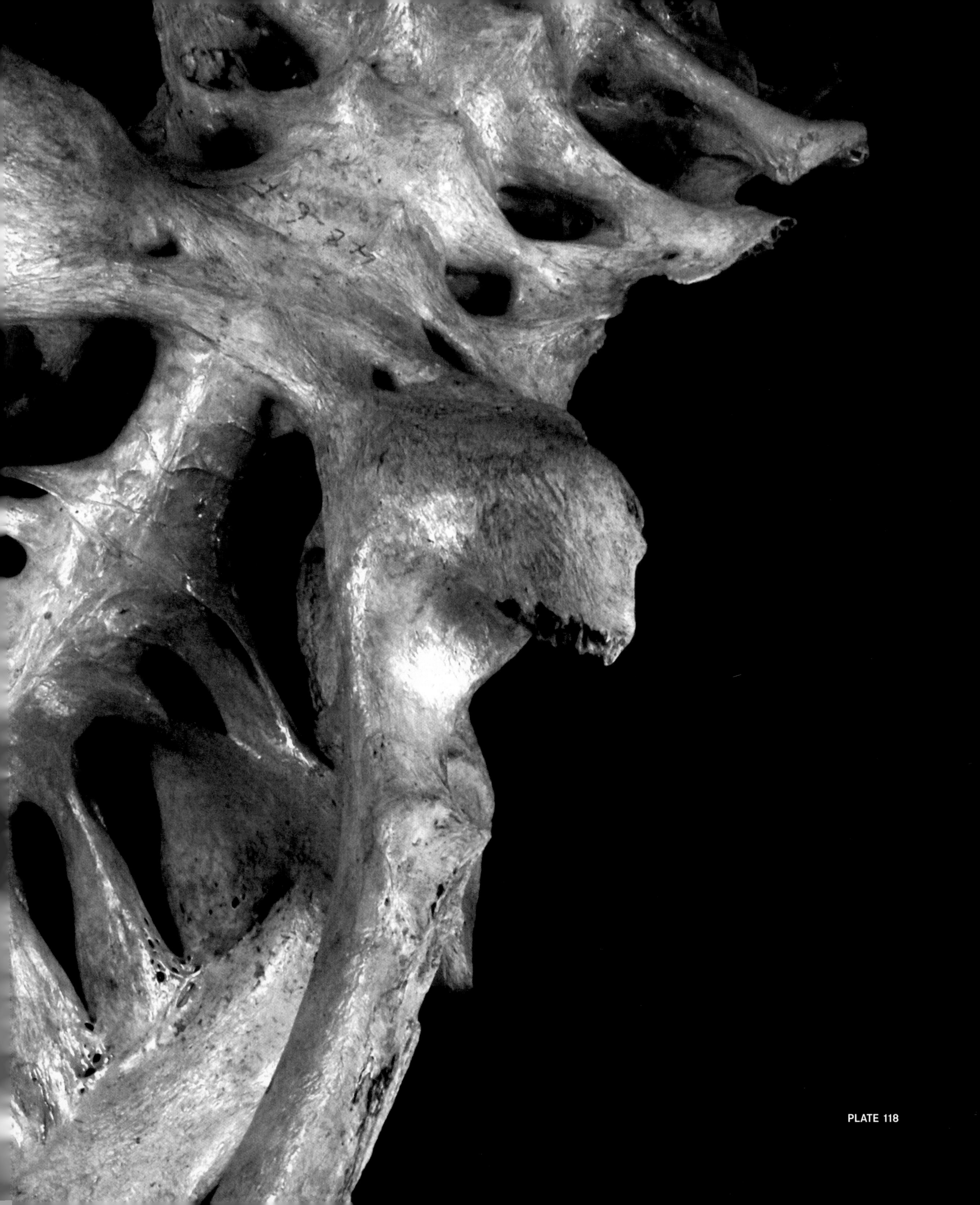

PLATE 118

But it is all too easy to go overboard here, and take all the blame for late Pleistocene extinction on our own shoulders. Many other species of plants and animals, including birds and small mammals that would not likely have been the focus of zealous hunting parties, also became extinct in the recent past. There is good evidence that many of the extensive habitat changes that accompanied the end of the last glacial advance were, in the manner that is by now familiar to us, perhaps the strongest contributor to the most recent wave of extinction. But there is that tantalizingly strong correlation between human arrival and the subsequent disappearance of many of the larger, more prominent mammalian (and even large bird) species. [Plate 118]

These issues are especially critical when we confront the current wave of extinction affecting our entire planet today. No one knows how many species of complex animal and plant life there were, say, in 1900. Some say ten million, some say perhaps as many as eighty million. But one thing is certain: however many species there were a century ago, there are considerably fewer now.

The invention of agriculture, in particular, has transformed our species in many ways. It has allowed a settled existence with a reliably steady source of food. With that has come a proliferation of roles that humans play in society. Agriculturally-based economies have allowed a flowering of civilization that includes the very finest cultural achievements. (Though we should not lose sight of the exquisite cave paintings skillfully executed 10,000 years before the development of settled communities dependent on agriculture!).

But agriculturally-based social systems quickly engendered rapid population growth. The ills for humans that come as a result are well known. We should consider here the effect that agriculture and our burgeoning population has had on the rest of the world over the last 10,000 years. As population grows, food resources are stretched, and there is inevitably an effort to expand the resource base. As the supply improves, the population again expands: there has been, as yet, little sign that increased production is distributed to improve the average quality of life. Rather, population grows to gobble up the expanded resource base. [Plate 119]

No longer do we live in small bands, well integrated and in harmony with local ecosystems. We are, instead, insulated from the environment, the natural world, to a remarkable degree. We are, by now, a truly global species, with an interlinked economy that indeed reaches around the world. Our species, *Homo sapiens*, interacts on a global basis with the entire global ecosystem. We think we are removed from nature—because we no longer need to take direct heed of our local ecosystems. But we will never really escape the natural world, as a growing number of people the world over are beginning to realize.

Our continuing expansion of agriculture and other forms of economic “development” are directly destroying habitat. As we have seen, it is habitat alteration that causes mass extinction. There are physical causes of habitat change that have caused mass extinction totally without our help in the past. The strongest, most consistent cause of habitat alteration has been global climate change. As we have seen for the K-T extinction, though, there can be additional factors that perturb the system—complicating it and making a bad situation worse.

I cannot help but think that *Homo sapiens* is the current version of late Cretaceous comets. In discussing the hunting hypothesis of Pleistocene extinction, I deliberately used the word “impact.” We are still in the throes of environmental change stemming from the last glacial melt-back. No one knows what is going to happen next to the climate. On the one hand, we ourselves have been creating a phase of “global warming” through our use of fossil fuels and the consequent release of carbon dioxide into the atmosphere. On the other hand, we are several thousand years overdue for the next glacial pulse: the climate figures to get colder again fairly soon. Climatic instability is a certainty for the near future. That means habitat disturbance and quite possibly extinction. We are the wild card in the deck, the loose cannon, the modern version, perhaps, of those Cretaceous comets. But we are sentient beings, not mindless extraterrestrial bodies set on some rigid celestial trajectory. Maybe, just maybe, we can see what we are doing, and take steps to mitigate our growing negative impact on the world’s ecosystems. It matters, simply because we are a part (much as we seem to deny it) of the modern global ecosystem. True, extinction is “natural.” And, true, evolution seems to a very great degree to depend on extinction—as we are just about to see in the next chapter. But who cares about future evolution if we ourselves face disaster as the rest of our fellow species drop over the extinction cliff, changing forever the complexion of the ecosystems from which we ourselves sprang?

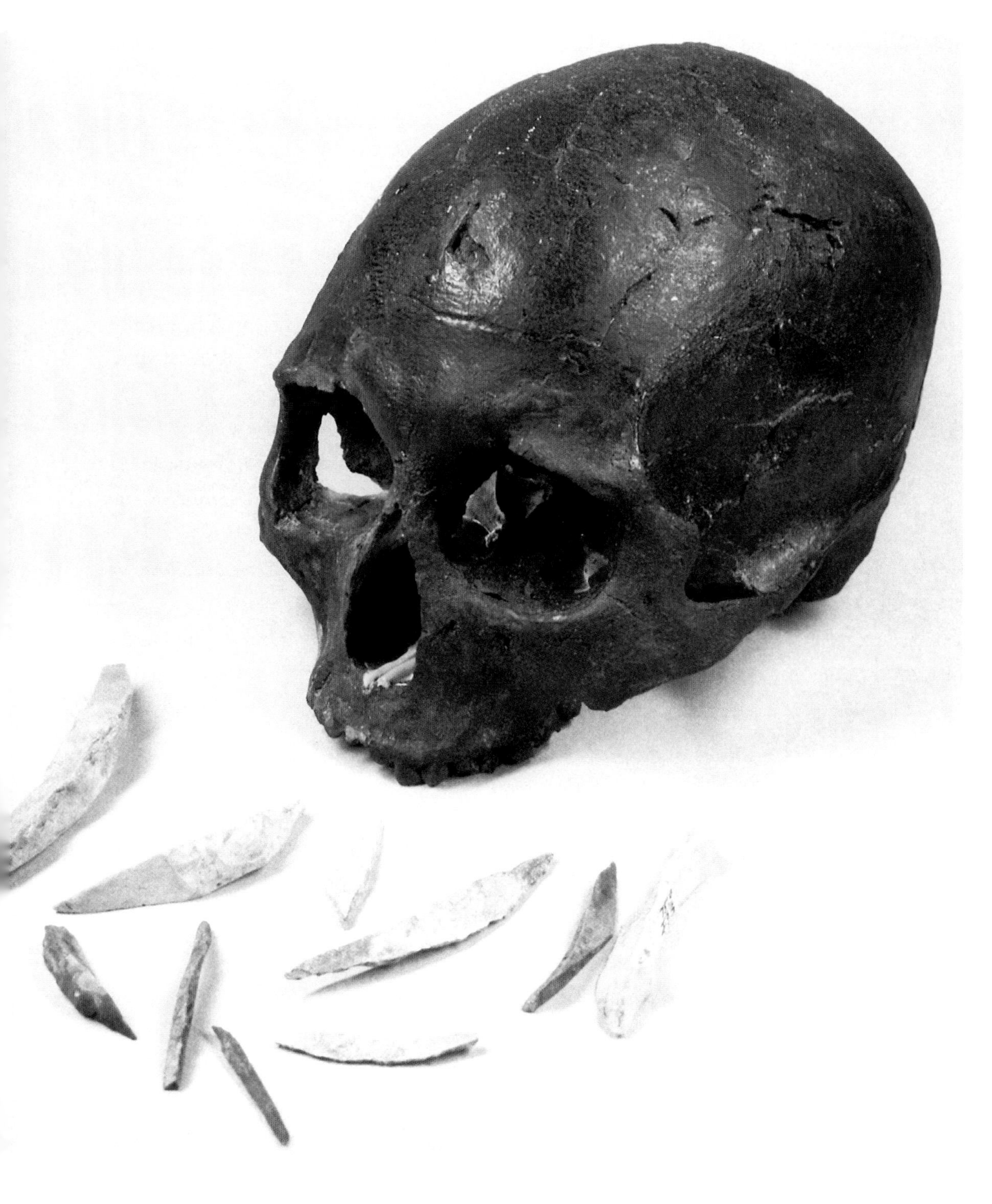

PLATE 119

NEOLITHIC MICROLITHS (STONE TOOLS) WITH A SKULL OF *HOMO SAPIENS*.

Tools from France, (*ca.* 10,000 years); skull from Chile (*ca.* 300 A.D.).

Agriculture, supplemented recently by the effluvia of industrialized society, is the major form of habitat alteration from human activity.

CHAPTER 7

MACRO–EVOLUTION

PLATE 121

***PROTOROHIPPUS VENTICOLUS*, A SOCALLED DAWN HORSE.**

Lower Eocene (*ca.* 52 million years). Fossil Lake, Green River Formation, Wyoming.

Paleontologists were surprised to learn that, although horses were known only from the Old World in the modern Era, and had to be imported back to the Americas, they actually did most of their evolving in North America. This is one of the earliest, belonging to the famous "dawn horse" brigade. (See Plate 133 – 135.)

Credit: National Park Service.

Egypt, at first glance, would seem to be a most unpromising spot to look for fossil whales. But Earth's habitats are constantly shifting around: 10,000 years ago, what is now the barren Sahara was verdant. Fifty million years ago it was under water. The cliffs that line the Nile Valley from Cairo south almost to Aswan are loaded with Eocene fossils: clams, snails, crabs, sand dollars and the gigantic one-celled shelled amoebae known as "nummulites" ("coin-stones" for their flat, round coin-like shape). Not far to the southwest, across the Nile from modern Cairo, lies the Fayyum Depression with Birkat Qarun, Egypt's largest natural lake. Just north of the lake, the most precious of Egypt's many fossil treasure troves stands out from the desert surroundings. The Fayyum beds are especially rich in fossil vertebrates. The upper layers are particularly important because they house the earliest known elephants (*Phiomia* and *Moeritherium*) [Plate 121], some unique and rather strange beasts known nowhere else (for example, the superficially rhino-like *Arsinoitherium* [Plate 122], named for an ancient Egyptian queen), as well as the earliest known species of higher primates. But it is the marine beds below that hold our attention here: that's where bones of some of the world's oldest whales come from. And whales are as good a place to begin a contemplation of *macroevolution* (meaning large-scale evolution) as any I know.

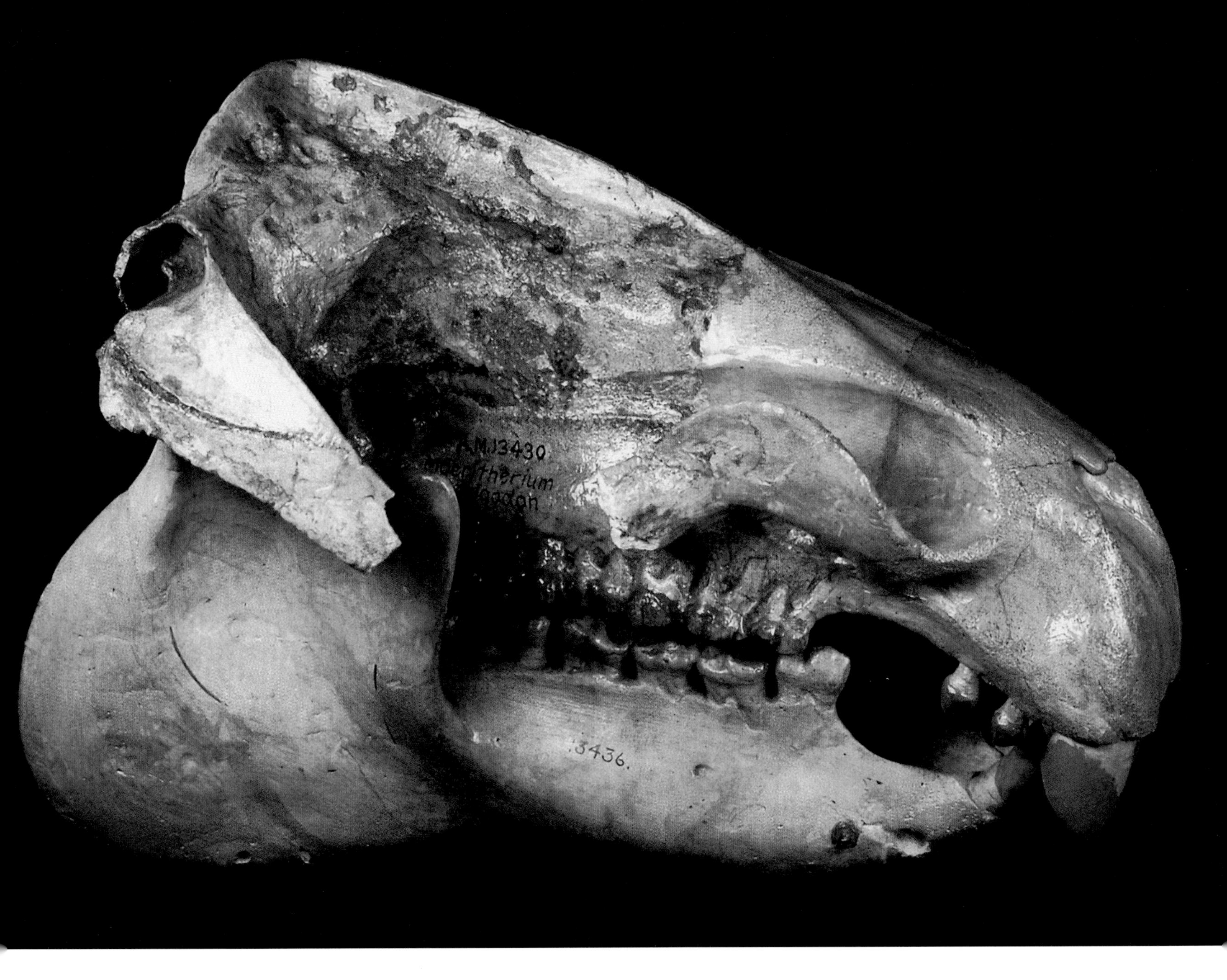

PLATE 121

***MOERITHERIUM TRIGODON*, A PRIMITIVE ELEPHANT.**

Lower Oligocene (*ca.* 35 million years). Fayyum, Egypt.

Despite their size and (therefore) long generation time, elephants typically evolve more rapidly than many smaller mammals.

PLATE 122

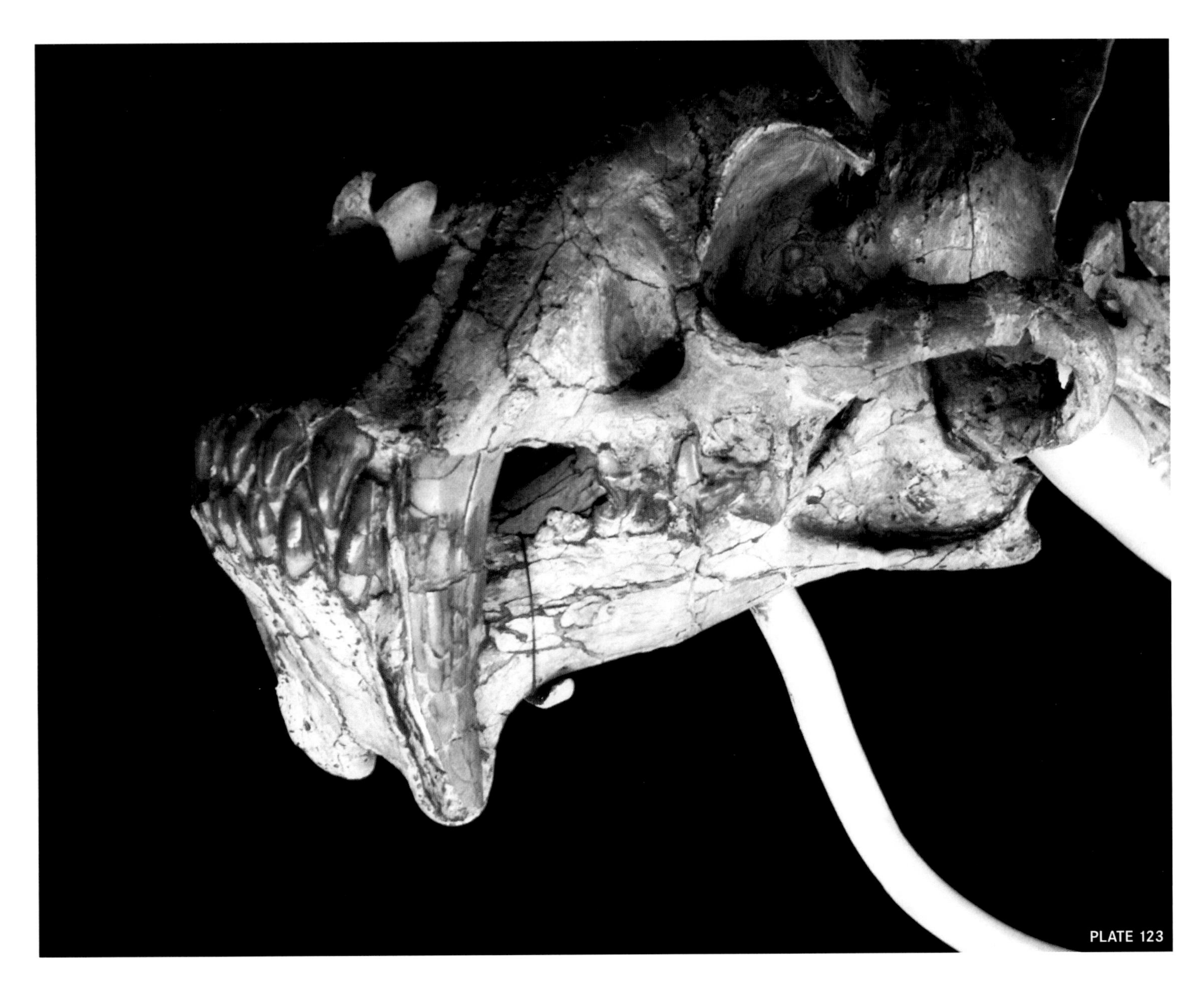

PLATE 122

***ARSINOITHERIUM ZITTELI*, A RHINO-SIZED PRIMITIVE MAMMAL.**

Lower Oligocene (*ca*. 35 million years). Fayyum, Egypt.

Along with the earliest members or well-known lineages such as whales and elephants, Egypt's fossil beds also reveal their share of exotic, and totally extinct, species whose evolutionary connections are not well understood.

PLATE 123

***HOPLOPHONEUS PRIMAEVUS*, AN EARLY SABER-TOOTHED "PALEOFELID" RELATED TO TRUE CATS.**

Oligocene (*ca*. 30 million years). South Dakota.

Darwin realized that, over geological time, evolution can produce a wide variety of organismic form. Macroevolution is the study of large-scale patterns of evolutionary change in geological time.

Whales, of course, are mammals. But they are very peculiar mammals: they have elongated, cigar-shaped bodies, front legs modified into flippers and no hind legs at all. Their internal anatomy and physiology has been extensively modified to permit life continuously immersed in oceanic waters, often for prolonged periods at great depths. Yet, like all other mammals, they derive their oxygen from the air. They have four-chambered hearts, hair, mammary glands, placentas and three middle ear bones, ample evidence that, however fish-like they may seem, their pedigree is definitely mammalian. And mammals are primordially creatures of the terrestrial world. Whales must have evolved from some terrestrial precursor. The major modifications required for re-adopted life in the sea neatly qualify as large-scale evolutionary change: whale evolution is an excellent example of "macroevolution." [Plate 123]

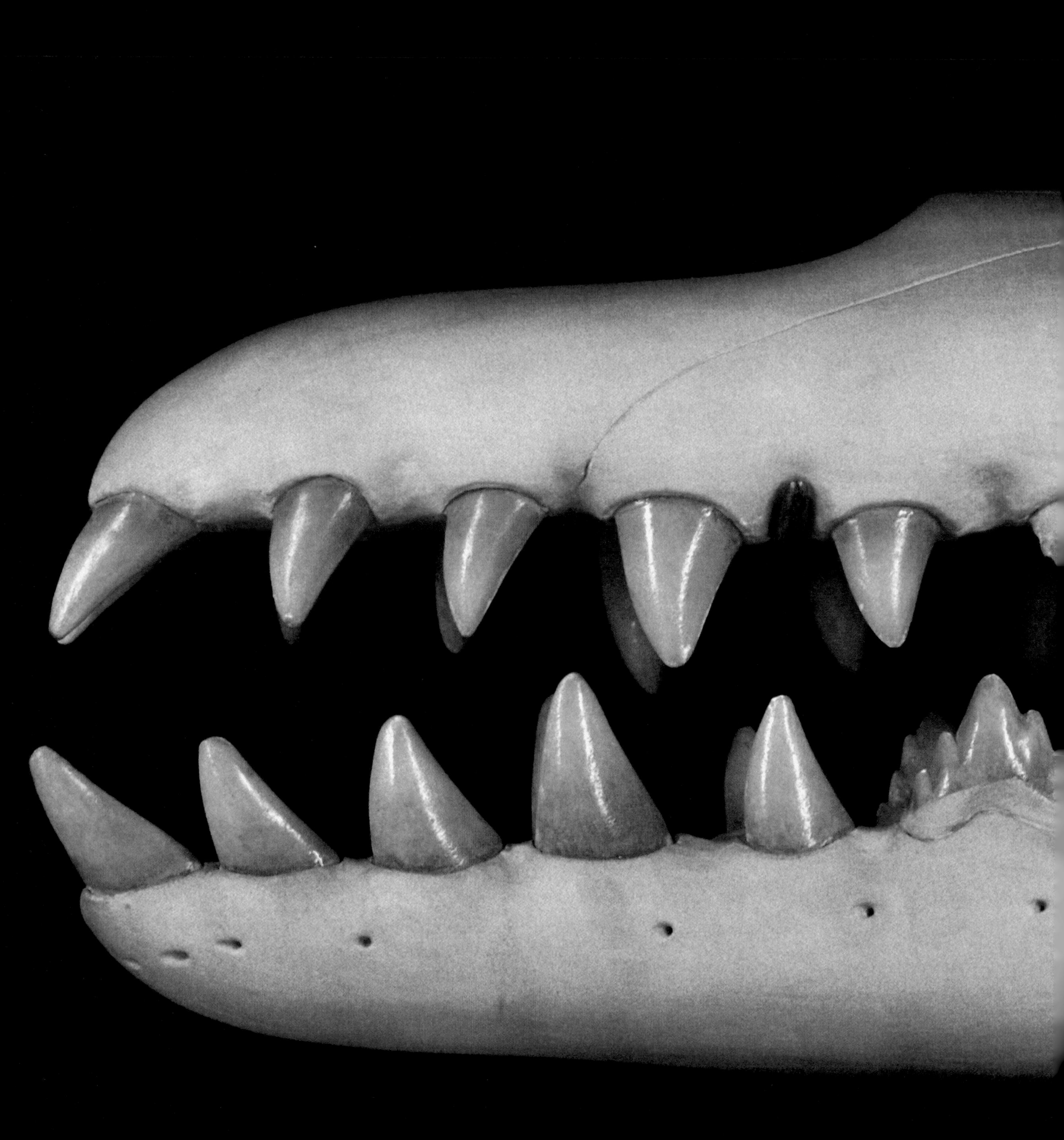

PLATE 124

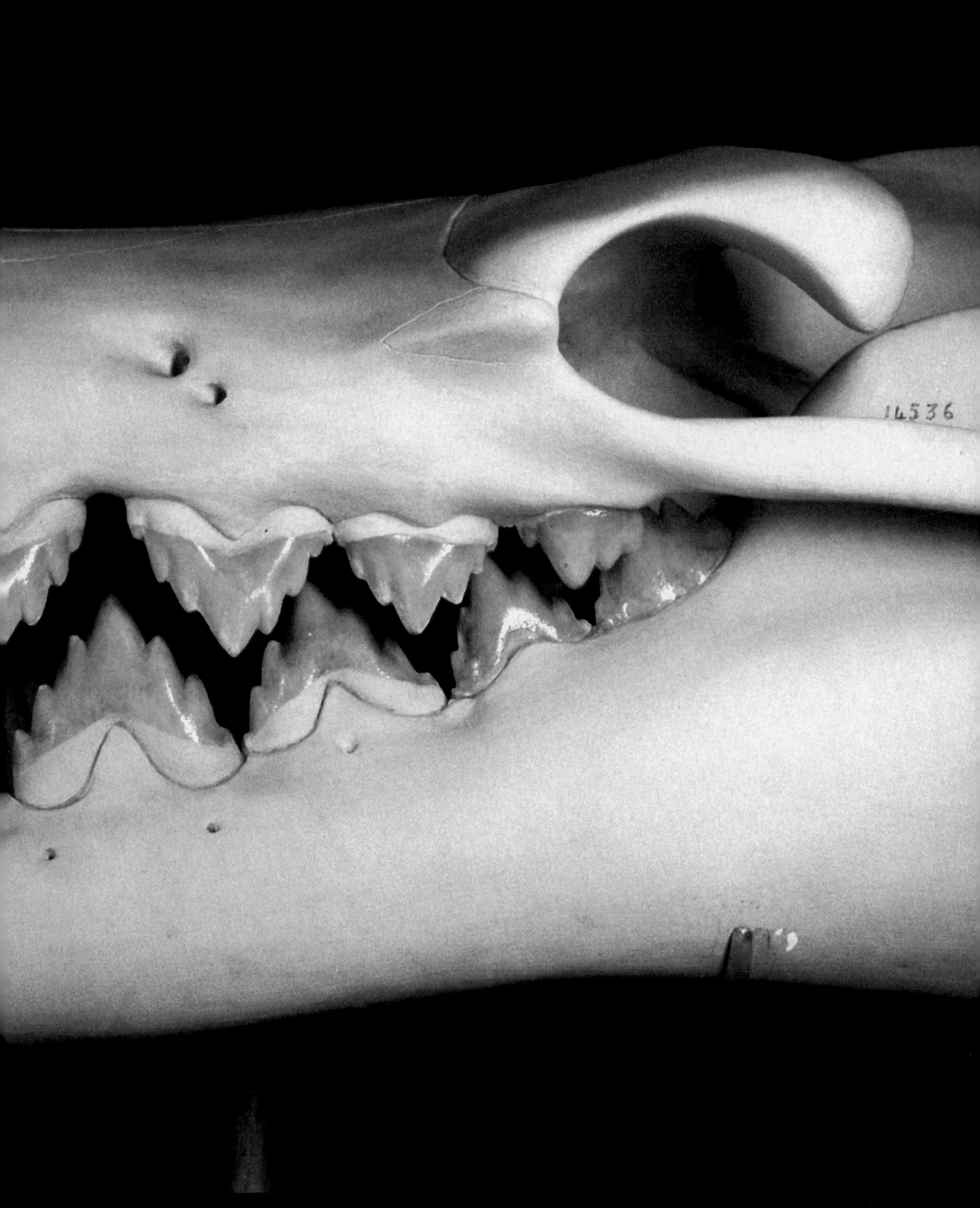
14536

PLATE 125

PLATE 124 (See page 198–199.)

***PROZEUGLODON SP.*, RECONSTRUCTED SKULL OF A PRIMITIVE WHALE.**

Upper Eocene (*ca*. 38 million years). Fayyum, Egypt.

Primitive for whales, nonetheless this species is unmistakably a true whale.

PLATE 125

***ANDREWSARCHUS MONGOLIENSIS,* A PRIMITIVE LAND MAMMAL CLOSELY RELATED TO WHALES.**

Middle Eocene (*ca*. 45 million years). Mongolia.

Though recent finds are closing the gap, the vast majority of primitive whales so far still resemble their advanced descendants far more closely than known fossils of terrestrial mammals.

But anyone looking at 50 million year old Eocene whale fossils from the Fayyum will immediately be struck by the fact that these early whales are very whale-like indeed. To be sure, they have features that seem quite primitive by modern whale standards: all Eocene whales have teeth that look more like terrestrial carnivore teeth than the simple, conical teeth of today's sperm whales. [Plate 124] And baleen whales, which lack teeth, and instead strain minute sea creatures with flat stringers of horny "baleen," had not yet put in an appearance in Eocene times. Some Eocene whales also retain vestiges of rear limbs as well. But in most essentials, Eocene whales are very definitely whales and not some form of creature distinctly intermediate between a terrestrial progenitor and full-blown whales. [Plate 125]

George Simpson led the greater part of his professional life pursuing problems in mammalian evolution at the American Museum of Natural History. It was Simpson's goal, in the late 1930s, to put together newly emerging ideas on the genetics of the evolutionary process with the large-scale patterns of evolution revealed by the fossil record. And the sort of evolutionary pattern provided by whales intrigued him deeply. Rather than blaming gaps in the fossil record for the absence of intermediate, transitional forms between the terrestrial ancestor and the Eocene whales, Simpson saw quite another message: Simpson realized that the fossil record of whales had something important to say about the very nature of the evolutionary process. He calculated that, if one took the conventional Darwinian position *i.e.* that evolution is generally even-handed, steady, slow, gradual and progressive and applied it to a case like the whales, the results were absurd. One can measure the average rate of evolution for various anatomical features, in the 50 million years it took to modify Eocene whales into fully modern forms. Let us then take that measured rate of evolution *within* whales and calculate how long it would have taken for Eocene whales to evolve from terrestrial ancestors. Extrapolating back, it would have taken at least 100 million years (possibly even considerably more) for the transition from terrestrial ancestor to aquatic, primitive whale descendant to have occurred assuming, that is, that whales evolved from terrestrial ancestors at the same rate of evolution we see in the 50 million years that elapsed between Eocene and modern whales.

Patently absurd, Simpson realized. One hundred million years prior to the Eocene puts us back into the Middle Mesozoic, long before true mammals show up as small, very primitive terrestrial creatures much like our modern "insectivores" shrews, hedgehogs and the like. [Plate 126] Whales, in fact, evolved from such a more advanced terrestrial mammal perhaps one of the "anthracotheres,"a group related to hippos as well as, presumably whales.

Simpson concluded that we cannot blame the lack of transitional forms on a faulty fossil record (Modern exploration has in fact revealed more details of the transition between terrestrial mammals and true whales.). The only sensible conclusion, Simpson realized, is that evolution, especially episodes of large-scale, truly *macroevolution,* must occur much more rapidly than the gentler pace typical of subsequent evolutionary transformation that takes place after a group becomes well established.

Nor, Simpson also saw, were whales an isolated example. Bats, too, tell the same sort of story to a mammalian paleontologist: the earliest bats, also Eocene in age, are primitive as bats, but recognizably and distinctively bats in all essentials nonetheless. [Plate 127] Chiropterist Nancy Simmons has found that the best preserved and one of the earliest Eocene fossil bats–*Onychonycteris finneyi* dated at *ca*. 52.5 million years–already had the complicated ear adaptations necessary for echolocation–the form of natural sonar bats use to navigate and catch insects on the wing in the dark. Bats, too, must have evolved rather rapidly from small bodied terrestrial hoofed animals that lived around the very end of the Cretaceous, related to whales, ungulates, and carnivores.

PLATE 126

***DARWINIUS MASILLAE*, A LEMUR-LIKE PRIMATE.**

Middle Eocene (*ca*. 47 million years). Messel Shales, Eichstatt, Bavaria, Germany.

Curiosity has always peaked when it comes to the evolution of our own mammalian Order Primates. Here is another exquisitely preserved fossil from the Messel Shales, which sheds light on the transition between the older forms of primates and the higher primates: monkeys, apes and humans.

Credit: Jørn Harald Hurum, University of Oslo.

PLATE 127

***ICARONYCTERIS INDEX,* A PRIMITIVE BAT.**

Lower Eocene (*ca.* 52 million years). Wyoming.

Primitive bats are fully bat-like in all essentials—and far removed from four legged, terrestrial ancestral mammals.

It isn't just mammals that tell this story. Virtually all the major groups of animals and plants show the same fossil pattern: rather abrupt first appearances in the fossil record, in a form that is destined not to change too radically throughout tens of millions of years during the rest of their recorded history. It is the large-scale, mega-version of the same sort of pattern that we see all the way down to the species level. New species appear relatively abruptly, implying rapid transition with little hope for finding samples intermediate between ancestral and descendant species. The origin of a species is typically followed by a vastly longer period of stability, or *stasis*, with little further evolutionary transformation: the phenomenon of punctuated equilibria. Now we see that this pattern is even more general: virtually all known instances of major adaptive change that characterize the origin of large groups of organisms (such as families, orders, or classes) occur quite quickly—especially when compared with the long subsequent histories of these groups where rates of evolutionary change are invariably much more modest.

How can we explain this phenomenon of rapid, almost explosive origin of adaptive innovation, this rapid rise of distinctive new groups? Once again, we ask: do these patterns mean we must abandon Darwin's vision of adaptive change through natural selection? Some evolutionary biologists ("*saltationists*") have thought so. Simpson, on the other hand, thought not: we need only to modify elements of the Darwinian vision to take into account the actual patterns of evolutionary change revealed in the fossil record.

MACROEVOLUTION AND THE QUESTION OF INTERMEDIATE FORMS

Rapid change by no means presupposes that natural selection does not produce an array of intermediate forms between the ancestral and descendant conditions. But rapid evolution, which is most likely to occur in small and rather restricted populations, will tend to reduce the chances of finding any traces of the intermediates in the fossil record. In punctuated equilibria, we are most likely to encounter fossils left by stable and far-flung species sometime during their long history; it is far less likely that we will be able to find specimens that were part of the small, localized populations living during the actual transitional speciation event itself. Much the same holds true for the rapid origin of larger groups of organisms. We can find fossil small bodied terrestrial hoofed animals and even fossil bats (though they are not at all common). Of the intermediates between these early Paleocene animals and bats, we have not a trace.

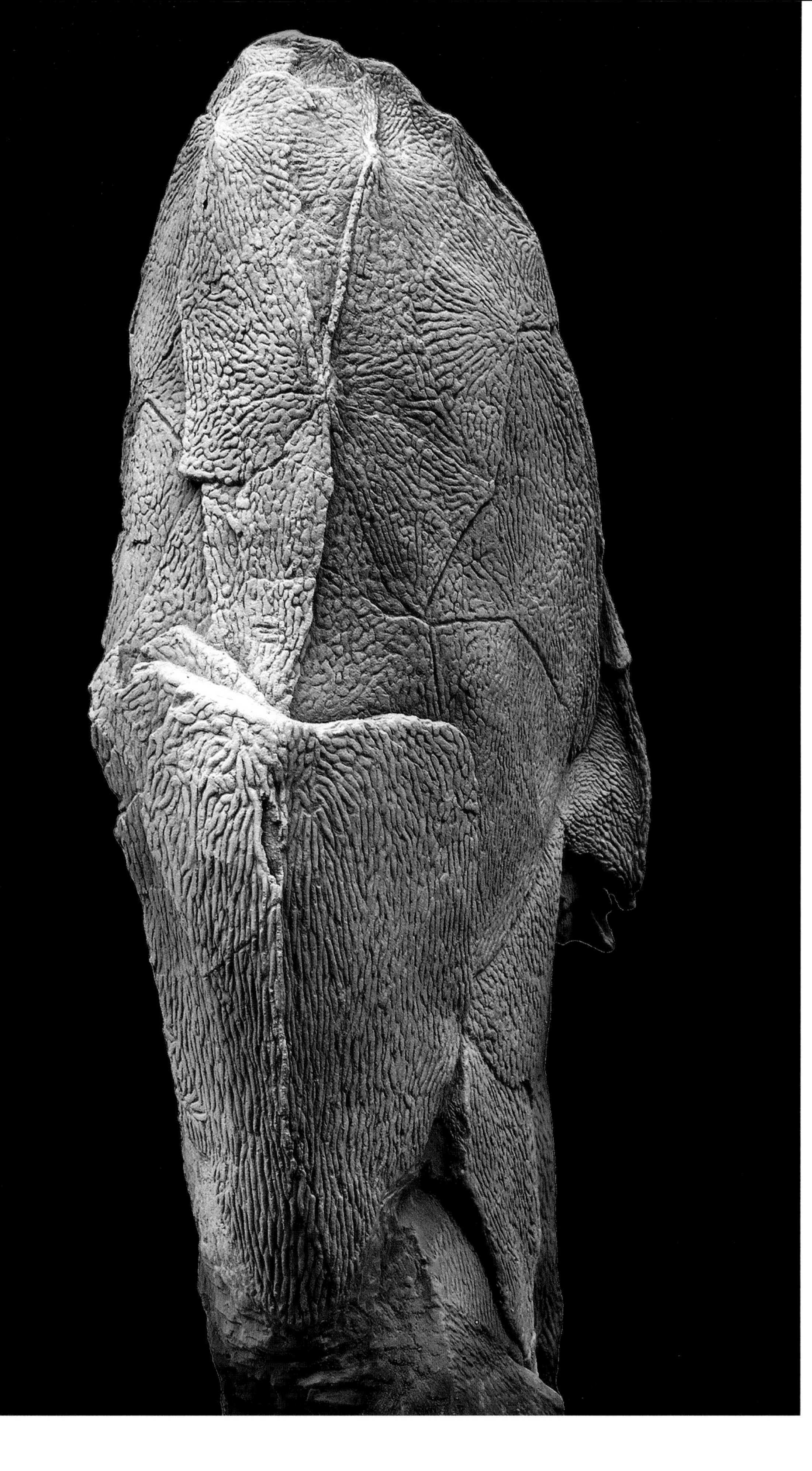

PLATE 128

ENTELOGNATHUS PRIMORDIALIS,
A JAWED PLACODERM.

Upper Silurian (*ca*. 419 million years). Kuanti Formation, Yunnan Province, People's Republic of China.

Like other placoderms, *Entelognathus* had armored plates, but it also had more advanced jaw bones providing evidence for the evolutionary link between placoderms and bony fishes. (See Plate 36.)

Credit: Zhu Min, Institute of Vertebrate Paleontology and Paleoanthropology.

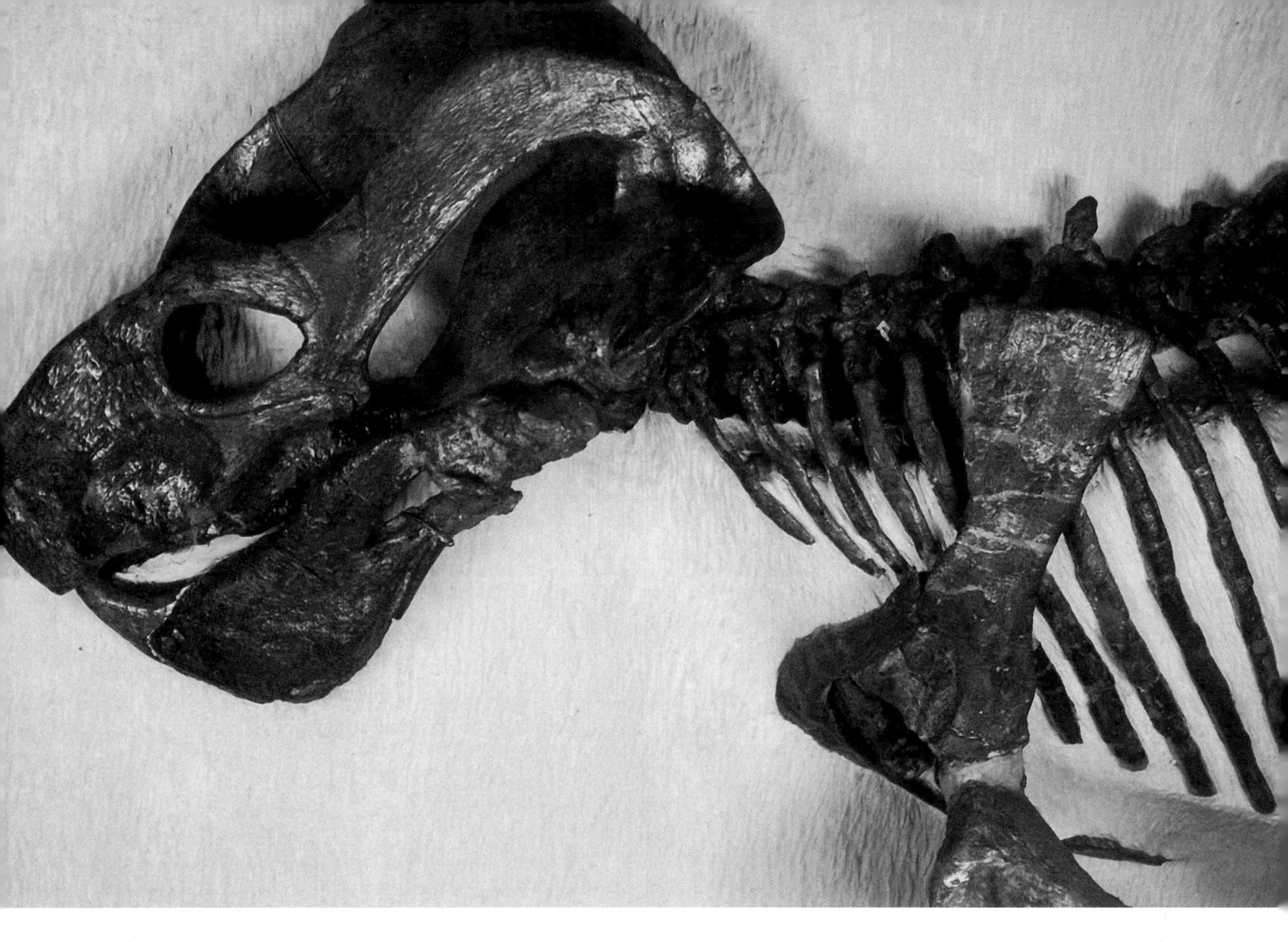

PLATE 129

ENDOTHIODON ANGUSTICEPS,
A DICYNODONT (PRIMITIVE MAMMAL-LIKE REPTILE).

Upper Permian (*ca*. 247 million years). South Africa.

The jaw joint of this species conforms to the old reptilian pattern—and there is but one bone (the stapes) in the middle ear.

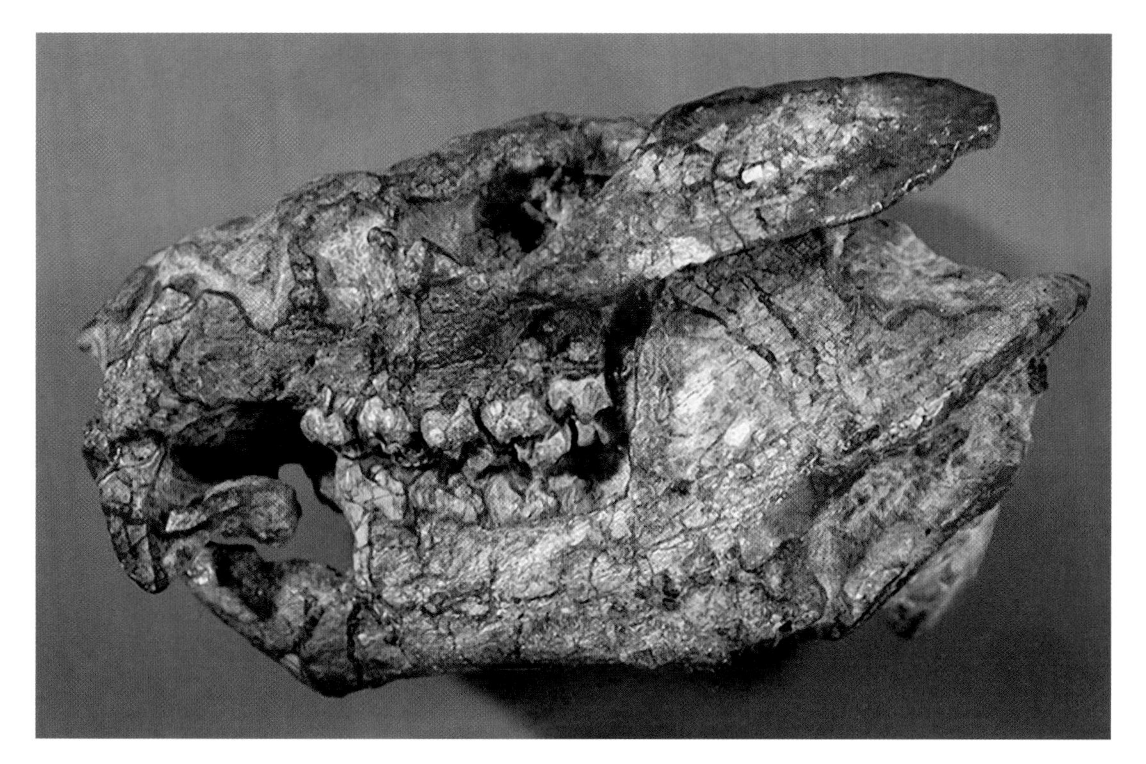

PLATE 130

***KAYENTATHERIUM WELLESI,* A TRITYLODONT (ADVANCED MAMMAL-LIKE REPTILE).**

Lower Jurassic (*ca.* 208 million years). Kayenta Formation, Glen Canyon, Arizona.

Tritylodonts, while retaining the old reptilian jaw elements in reduced form, show great expansion of the dentary bone, and appear close to the transition between reptilian and mammalian conditions.

Credit: Patricia Hol, University of California Museum of Paleontology.

But intermediates there were, and sometimes we are lucky enough to find them. [Plate 128] When we do, they amply validate a basically Darwinian interpretation of large-scale evolutionary change. In the Karroo sequence of present-day South Africa, for example, is a rich series of Permian and Triassic fossils that show a sequence of progressively more mammalian-like reptiles. [Plates 129, 130] The hinge between the jaw and skull in a mammal is formed by two bones: the "dentary" of the jaw and the "squamosal" of the skull. The hinge is formed by the "quadrate" and "articular" bones in the reptilian skull. Some of the South African fossils actually show a double jaw articulation: here is *both* the old reptilian quadrate-articular, as well as the newly evolved mammalian dentary-squamosal contact in the jaw hinge. Later, as the dentary-squamosal took over as the sole form of jaw joint in true mammals, the quadrate and articular became the incus and malleus—tiny bones which, along with the stapes already present in reptiles, form the characteristic three middle ear bones of all mammals. The fossil record does not record the details of the final transformation of jaw joint to middle ear bones—but the transition is still carried out in the embryological development of each and every living mammal!

Archaeopteryx, one of the earliest known birds, is also a prime example of an anatomical intermediate between major groups: in this case, terrestrial reptiles (carnivorous dinosaurs, in fact) and modern birds. Middle Jurassic in age, only five specimens have been recovered—all from the "Solnhofen" limestone of Bavaria (southern Germany). [Plate 131] The limestone is so fine-grained, in fact, that it was quarried initially to provide slabs for use in lithography. Able to transmit fine details of art work to the printed page, the Solnhofen limestone also beautifully reveals astonishing anatomical details from a by-gone era: around the bones we can see the fine impressions of feathers—the *sine qua non*, of course, of birds.

What do we really mean by "intermediates" in evolution? We might suppose that a creature intermediate between a bipedal carnivorous dinosaur and a house sparrow would somehow be intermediate in all respects. That's not what we see at all: *Archaeopteryx* has some advanced features of birds, and some primitive retentions—anatomical features held over from its reptilian ancestry. *Archaeopteryx* has the wings and feathers of a bird, but the face and tail of a carnivorous dinosaur. *Archaeopteryx* is a melange of reptilian and bird features. It is a *mosaic* rather than a blend, according to Sir Gavin deBeer, famed embryologist-turned-temporary-paleontologist in his famous monograph on *Archaeopteryx* written in the 1950s.

The basic Darwinian theory of adaptive match of organism to environment through natural selection remains intact. But clearly the original picture, one that predicted large-scale change as simply the accumulation of small incremental steps of change over long periods of time, does not work. [Plates 132,133,134] Simpson understood that the fossil record stands at odds with this simple picture: evolutionary rates are by no means constant. But the fossil record has even more to tell us about the circumstances surrounding major adaptive change, another key element of pattern in the history of life which is crucial to understanding how major adaptive transformation does in fact take place in the evolutionary process.

PLATE 131

ARCHAEOPTERYX LITHOGRAPHICA,
THE EARLIEST BIRD YET AUTHENTICATED.

Upper Jurassic (*ca.* 150 million years). Messel Shales, Eichstatt, Bavaria.

This is the famous "intermediate" between reptiles and true birds, a moment of evolutionary history captured by only seven specimens known so far.

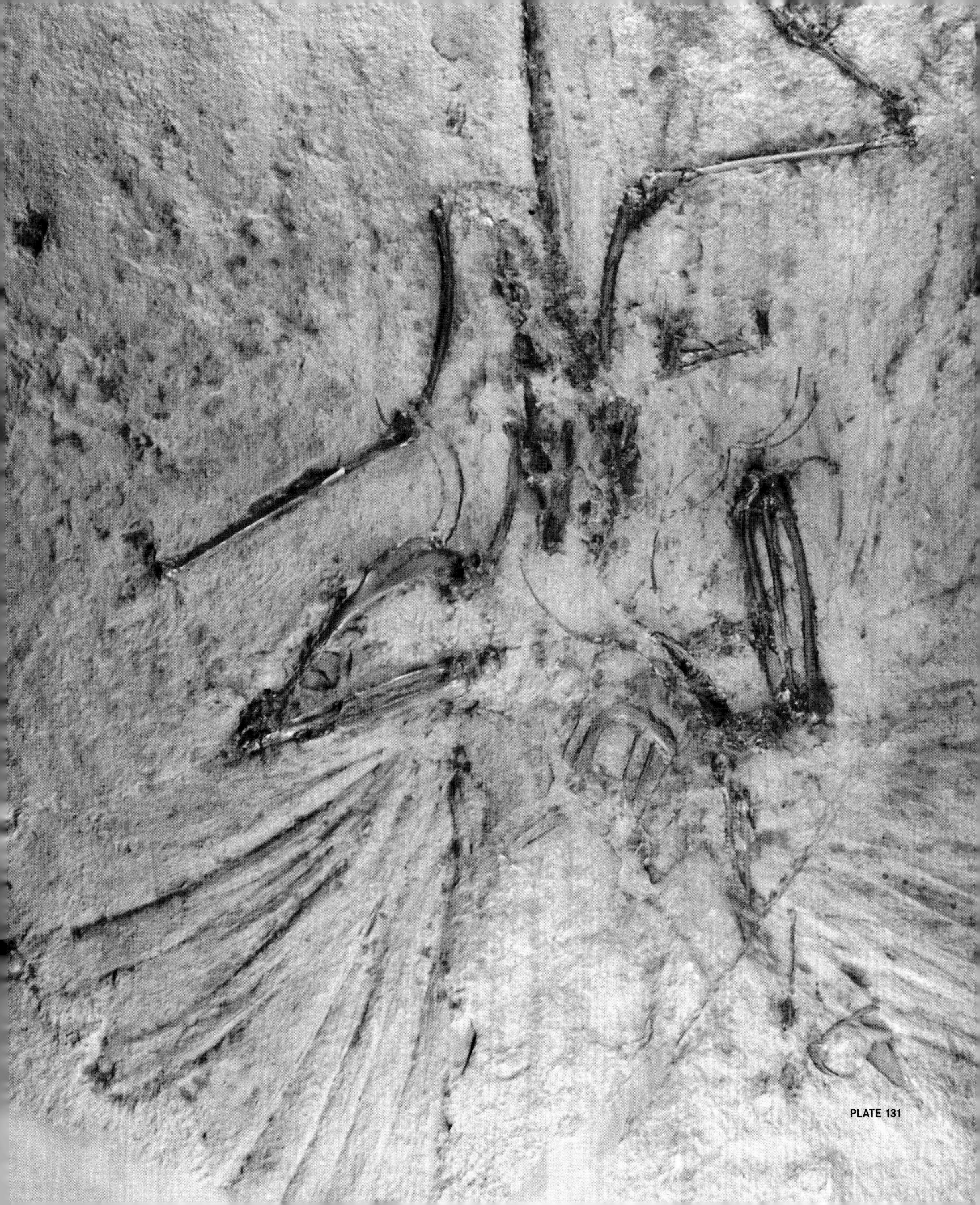

PLATE 131

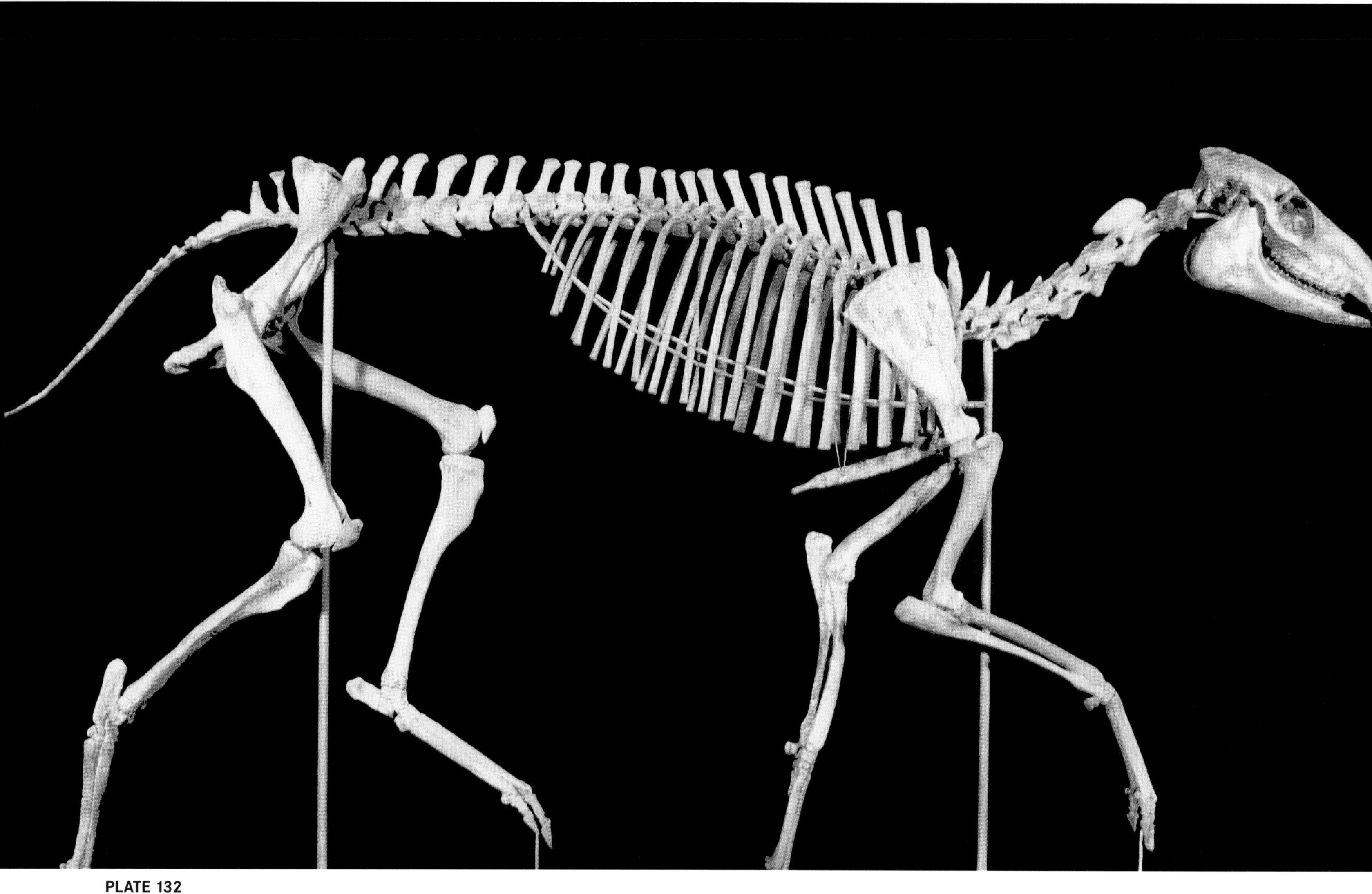

PLATE 132

PLATE 132

MESOHIPPUS BAIRDII,
A SPECIES OF PRIMITIVE HORSE.

Oligocene (*ca*. 30 million years). South Dakota.

Even though most of horse evolution took place in North America, early studies of fossil horse relatives in Europe were among the first to apply Darwinian principles to the fossil record. (See Plate 120.)

PLATE 133

MESOHIPPUS BAIRDII, **ANOTHER VIEW OF THE SAME SPECIMEN AS PLATE 132.**

Evolutionary changes in relative size of the body, number of toes on the feet and conformation of teeth seem at first glance to have been gradual and progressive in horse evolution. However, George Simpson pointed out that horse evolution was not simply a matter of the modification of traits within a single lineage, but instead involved continual diversification of different species from the Eocene right up to modern times.

PLATE 134

MESOHIPPUS BAIRDII,
CLOSE-UP VIEW OF SAME SPECIMEN AS IN PLATE 132.

Simpson utilized horse evolution as an example of his idea of "quantum evolution"—where the change from browsing to grazing was considered to be a very rapid, "all or nothing" shift in adaptive properties.

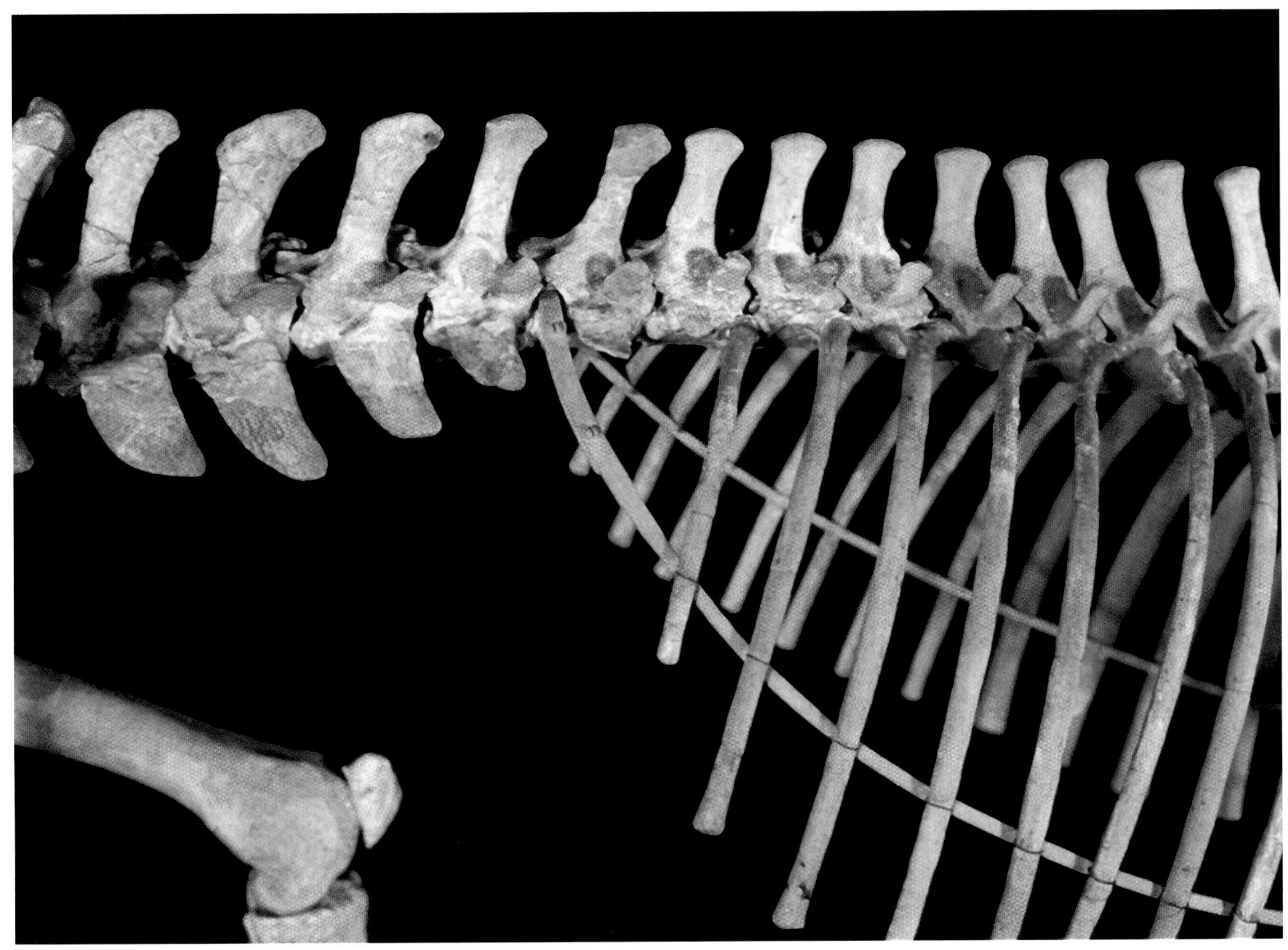

PLATE 133

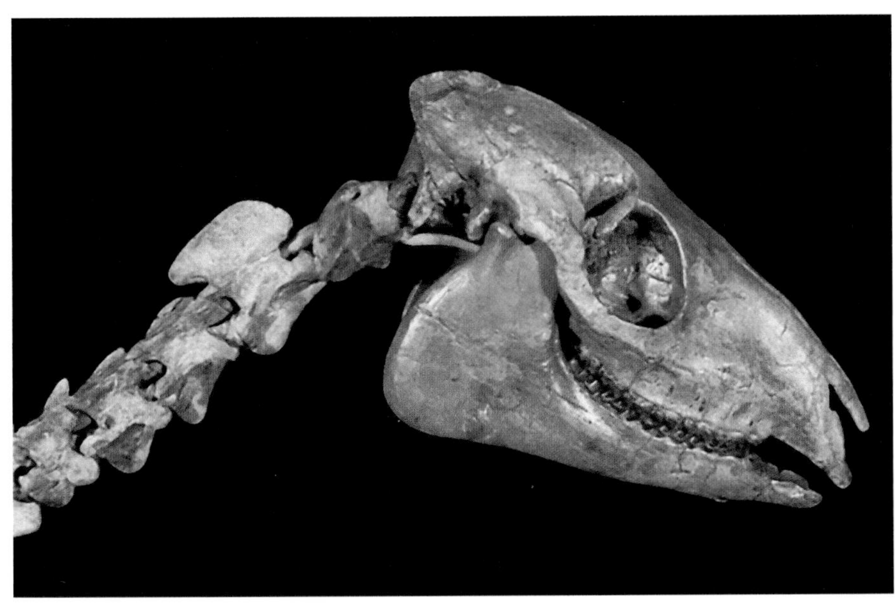

PLATE 134

PLATE 135 (See page 212–213.)

UNIDENTIFIED FORM OF A STROMATOLITE, A LITHIFIED BLUE-GREEN ALGA.

Precambrian (*ca.* 1.5 billion years). Death Valley, California.

This stromatolite was formed well after the advent of multicellular life forms.

PLATE 135

MACROEVOLUTIONARY PATTERN

Nature truly abhors a vacuum. Evolutionary paleontologists of the modern era are convinced this old maxim is as true of the evolutionary process as it is in any other sphere of natural phenomena. On a small scale, if there is the requisite variation, natural selection will quickly modify the traits of organisms to match slightly altered conditions. On the ecological scale, if new habitats open up, organisms will quickly arrive and begin taking advantage of the new opportunity. Volcanic islands are wonderful examples: when Krakatoa, between Java and Sumatra, erupted explosively in 1883, all its living creatures were destroyed. But plants and, soon thereafter, a varied assembly of animals quickly reappeared, coming in from the neighboring larger islands. Surtsey, an island which arose from the mid-Atlantic off the coast of Iceland as recently as 1966, likewise was quickly populated.

Very much the same sort of pattern occurs in macroevolution. And therein lies an important clue to understanding the dynamics of the history of life. The fossil record loudly proclaims the generalization that, when opportunity arises *and there are no (or only a very few) species already in existence* that are able to capitalize, new species will appear that can take advantage, species whose organisms can make a living and reproduce in the new circumstances. When vast areas open, evolution is rampant. And it is then that the larger-scale transformations take place.

Two sorts of circumstances present virgin evolutionary territory, the kind that is often quickly followed by major evolutionary change. One is simply living systems confronting, for the first time, entirely novel habitats. In some circumstances, individual lineages will enter environments already occupied by previous inhabitants. Birds and insects—not to mention the already extinct flying reptiles of the Mesozoic (the pterosaurs)—had already been flying long before bats took to the air. Likewise, different groups of marine reptiles (plesiosaurs and ichthyosaurs) had already taken the plunge back into the sea, evolving from terrestrial ancestors long before the mammalian whales followed suit.

These are isolated instances. What happens when entire new realms open up for the first time is something entirely else again. Such events mark some of the most profound milestones in the history of life. Especially in terms of the evolution of multi-cellular animals and plants, two episodes were especially critical. The first was the sudden radiation of complex animal life near the base of what we call the Cambrian Period some 570 million years ago. The second was the Silurian invasion of the land by many different lineages, starting with the evolution of higher plants, but including different arthropod lineages (insects, spiders and relatives), mollusks, annelid worms and, of course, vertebrates. The very first realm to support life was, of course, water: the ancient seas that have covered most of Earth's surface for billions of years. The evolution of all the major phyla took place beneath the waves. Fossils left behind by these ancient creatures show us that macroevolutionary patterns were established very early in life's history. [Plate 135]

The earliest fossils, simple bacteria, are about 3.5 billion years old. For life's first 1.5 billion years, bacteria appear to have been the only forms of life—though some of these simple-celled "prokaryotes" grew together in mat-like form, adding layer upon layer to form big cabbage-shaped mounds on the sea floor. [Plate 136] Complex single-celled organisms ("eukaryotes") show up in rocks dated at about 2 billion years. Eukaryotic cells are found in absolutely all forms of life except bacteria: algae and amoebae are single-celled eukaryotes. But plants, animals, and even fungi are multi-celled organisms with the same basic eukaryotic cell plan as a lowly single-celled amoeba. In the interim between the first appearance of prokaryotes and the arrival of eukaryotes some 2 billion years later, geochemical evidence suggests that Earth's atmosphere had finally gained enough oxygen (from outgassing of volcanoes, as well as the photo-synthetic activities of some of the early prokaryotes) to change from a reducing to an oxidizing medium. The vast majority of modern creatures rely directly on oxygen for the basis of their physiological activities.

PLATE 136

***CRYPTOZOON,* A STROMATOLITE.**

Upper Cambrian (*ca*. 515 million years). Saratoga Springs, New York.

Bacteria were the first, and for about 2 billion years, the only form of life on Earth (but see Plate 34).Stromatolites are still forming in such places as Shark Bay, Australia.

PLATE 137

DICKINSONIA SP.,
A WORM-LIKE METAZOAN.

Upper Precambrian (*ca.* 600 million years). Ediacara, Australia.

An extensive fauna of softbodied animals of debatable evolutionary affinities is found in rocks of uppermost Precambrian age.

We have little record of what conditions were like on the land areas exposed to the Precambrian atmosphere. But we do know that life arose in the sea and was for billions of years entirely confined to aqueous habitats. For at least 600 million years after complex celled life had appeared, life continued to be restricted to minute aquatic prokaryotic and eukaryotic organisms. Then, beginning perhaps 635 million years ago, traces of small shelled invertebrates begin to show up in the fossil record.

Some 20 million years before the Cambrian, we encounter the earliest undisputed traces of large complex multicellular life yet known. Remains of soft-bodied coral and worm-like forms, from such places as Australia's Ediacara Hills, record the oldest known phase of metazoan (complex animal) history. Whether they were the forerunners of the trilobites, brachiopods and other lineages that show up, abruptly, at the base of the Cambrian, is currently a bone (or at least a shell) of scientific contention among paleontologists. [Plate 137]

In recent years, just below the base of the Cambrian, careful field studies have revealed a wealth of minute shelled fossils, plus the tracks and trails of larger-bodied creatures, among them undoubtedly primitive trilobites. But then, relatively suddenly, at the beginning of the Cambrian Period, within a span of a mere 10 million years, all the major groups of complex animal life—*all the phyla*—appeared. Ten million years may seem like a vast stretch of time: by most criteria, it is a lot of time. But consider that nearly 3 billion years had already gone by since life had left its first traces in the fossil record. And consider, too, that no new animal phyla are known to have originated since the early Cambrian. [Plate 138]

PLATE 138

ONYCHODICTYON FEROX,
AN EARLY ANCESTOR OF LIVING VELVET WORMS.

Lower Cambrian (*ca.* 530 million years ago). Chengjiang County, Yunnan, People's Republic of China.

This bizarre looking marine animal has been shown to be related to the so called "velvet worms," known today as the terrestrial Phylum Onycophora. Onycophorans have long thought to bridge the evolutionary gap between annelid worms (Phylum Annelida) and arthropods (Phylum Arthropoda).

Credit: Jianni Liu, Northwest University.

Here again we find a familiar pattern—on a truly grand scale: relatively suddenly, the whole spectrum of invertebrate life—including sponges, brachiopods, arthropods (trilobites, chelicerates and crustaceans), mollusks, plus spineless chordates in the same phylum as the vertebrates—burst on the scene, the world over. By the end of the Cambrian we have records for all the major groups of hard-shelled invertebrate organisms—and some evidence that vertebrates had appeared as well (though the oldest definite fish fossils are Middle Ordovician in age). [Plate 139] What could have caused such a proliferation? Why did it happen at the point in time we retrospectively call the "base of Cambrian?" Suspicion continues to rest on the continued rise of atmospheric oxygen to the point where this vital substance rose to a critical level of saturation in sea water, high enough for the first time to enable the metabolic activities of large, complex animals. Oxygen tension in the water passed a critical threshold, creating an exploitable environment—and evolution produced its closest possible approximation of an immediate response. Others have proposed that a rise in sea level and consequent change of oceanic chemistry may have prompted the relatively sudden evolutionary burst of shelled animals—and the calcified skeletons typical of vertebrates.

PLATE 139

***PHYLLOGRAPTUS SP.,* A GRAPTOLITE.**

Lower Ordovician (*ca*. 485 million years). Idaho.

Floaters in the primordial seas, graptolites are extinct relatives of hemichordates—and thus distantly related to chordates, including ourselves.

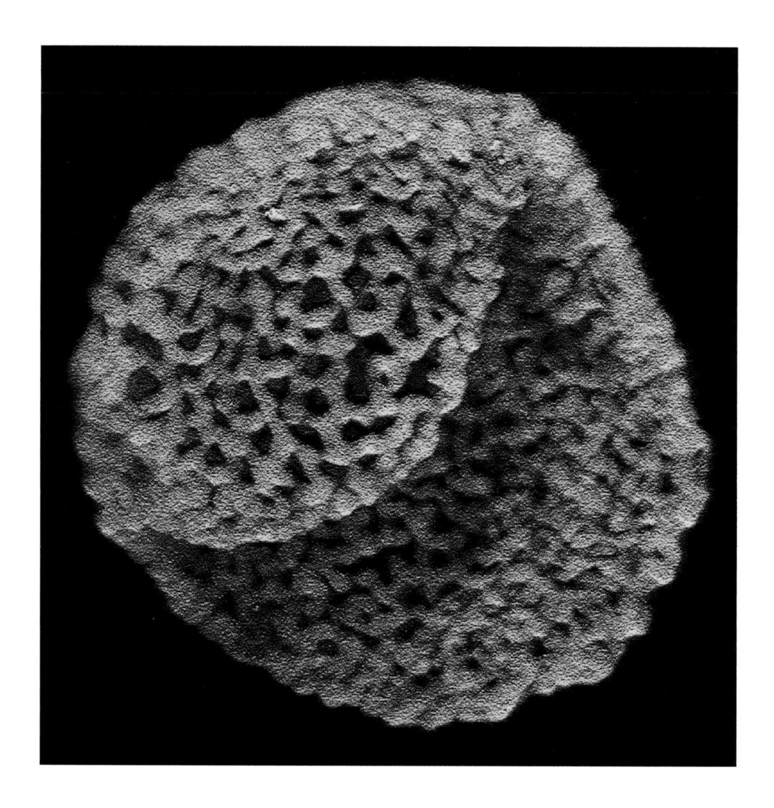

PLATE 140

FOSSILIZED ANGIOSPERM-LIKE POLLEN GRAIN, AN ANCESTRAL FLOWERING PLANT.

Early Triassic (*ca*. 250 million years). Germanic Basin, Northern Switzerland.

This fossil suggests that angiosperms—modern flowering plants—may have evolved in the early part of the Triassic, not long after the great Permo-Triassic extinction event.

Credit: Peter A. Hochuli, University of Zürich.

Opportunity knocks, the door is opened and life proliferates into forms that have never existed before. Much the same thing began to happen when life invaded the terrestrial environment, just a bit over 400 million years ago, during the Silurian Period. First primitive plants established an ecological beachhead. [Plate 140] They were followed swiftly by the first insects and other terrestrial arthropods, including an array of spider-like creatures. By Middle Devonian times, around 385 million years ago, there were many forms of complex plant life. Indeed the earliest forests, albeit of tree ferns, had already appeared. Crossopterygian fish, with their lobed fins and lungs that had not been converted to swim bladders (as they had in their close relatives, the actinopterygians, the primary group of fishes with us today) were preeminently suited to ever-longer forays out on land. They soon gave rise to the earliest amphibians.

But the Cambrian and Silurian explosions were exceptional circumstances. Most of the evolutionary history of lineages does not involve entering into previously unoccupied ecological realms. It is relatively easy to fathom how living forms, once conquering previously intractable physiological problems (such as insufficient oxygen availability, or the problem posed to aquatic organisms by conditions out on dry land), will proliferate into a wide variety of niches. But this cannot be the whole story of macroevolution: there has been too much evolutionary history, too much adaptive change through time that has not come as a simple invasion of wholly new ecological regimes. What really goes on once ecosystems are up and fully running? And how do the fates of ecosystems enter into the macroevolutionary process?

EXTINCTION AND MACROEVOLUTION

We have radically changed our perspective on dinosaurs in recent years. More paleontologists than ever are feverishly working away in the field and lab, learning more and more about these symbols of ancient life. By now, virtually everyone is aware that dinosaurs first appeared in the Upper Triassic and lasted for over 150 million years, only finally disappearing at the very end of the Cretaceous, during the most famous of the ancient mass extinction events. [Plate 141]

PLATE 141

VELOCIRAPTOR MONGOLIENSIS,
A CARNIVOROUS DINOSAUR.

Upper Cretaceous (*ca*. 90 million years). Flaming Cliffs, Mongolia.

Not all dinosaurs were large, as the skull of this little carnivore clearly shows.

PLATE 142

***ORNITHOLESTES BERMANNI,* A CARNIVOROUS DINOSAUR.**

Upper Jurassic (*ca.* 150 million years). Como Bluff, Wyoming.

Without the eventual extinction of dinosaurs, mammals would not have been able to diversify and supply their own players to the ecological arenas of life.

Dinosaurs were the major animal constituents of the terrestrial ecosystems of the Mesozoic. The herbivores among them came in all sizes; some lived in herds and may even have migrated with the seasons. There were carnivores large and small. Think of a modern tropical mammalian community, say in East Africa. The very same degree of ecological complexity was attained in many dinosaur communities of the Mesozoic. [Plate 142]

When wily King Kong fights the stupider, if stronger and fiercer, Godzilla in the movies, he depends on brains to combat superior brawn. Their fight is a fictionalized version of what, until very recently, was the prevailing interpretation of dinosaur extinction: dinosaurs were widely thought to have been out-maneuvered by the newly evolved mammals. Their time had come, and they could no longer out-compete their more adroit mammalian competitors. The mammals took over, after out competing their rivals—and the rest, as they say, is history.

That particular notion is, of course, by now as dead as the dinosaurs. Dinosaurs were active, able and ecologically well diversified. They suffered through many setbacks during the course of their existence, always coming back until they were finally done in at the end of the Cretaceous. Mammals, meanwhile, had been around just as long as the dinosaurs: both show up in the fossil record in the Upper Triassic. Early mammals, however, were relegated to the position of the "rats of the Mesozoic," filling small roles in the ecological communities dominated by their reptilian vertebrate relatives. [Plate 143]

The truth is that those “clever” little mammals would still be running around, dodging dinosaur feet and accomplishing little else, had not the dinosaurs finally succumbed—through no real fault of their own—during that major ecological collapse that brought so much of life to a crashing halt at the end of the Mesozoic, at the K-T boundary. The mammals managed to survive—though they too suffered losses during that big mass extinction. But it wasn’t until a very general and world-wide ecological collapse took out the dinosaurs (along with so many other kinds of animals and plants) that the mammals finally were faced with life on Earth *for the first time without dinosaurs to contend with*. [Plate 144]

PLATE 143

***EOMAIA SCANSORIA,* THE EARLIEST KNOWN PLACENTAL MAMMAL.**

Lower Cretaceous (*ca.* 125 million years). Yixian Formation, Liaoning Province, People’s Republic of China.

The discovery of this fossil shrew-like mammal places the apparent age of the origin of true (“eutherian” or “placental” mammals all the way back to the Lower Cretaceous.

Credit: ZheXi Luo, University of Chicago.

PLATE 144

BUXOLESTES PISCATOR,
AN OTTER-LIKE
PLACENTAL MAMMAL.

Middle Eocene (*ca.* 45 million years). Messel Shales, Eichstatt, Bavaria, Germany.

Species in this genus were part of the first mammal evolutionary radiation following the demise of the dinosaurs and other megareptiles.

Credit: Gabriele Gruber, Hessisches Landesmuseum Darmstadt.

And there, in a nutshell, is the significance of extinction to macroevolution. Time after time we see examples of groups that survive for long periods of time, usually with many different species (though, as we have seen, living fossils characteristically come from lineages that typically have very few species around at any given moment in geological time). Individual species are continually becoming extinct, and new ones are evolving, but nothing major ever happens in terms of evolutionary innovation within the group as a whole. Then the group is cut back drastically by an episode of extinction. If some of its members happen to squeak through, there is typically a period of proliferation—often after a bit of lag: it takes time for life to recover, and for speciation to produce all the new players for the ecological arena. And that's when innovation typically shows up.

Ammonoid cephalopods are a compelling case in point. Arising from nautiloid-like ancestors early in the Devonian Period, these shelled relatives of squid and octopi had a long and exuberant history. Proliferating greatly in the Upper Paleozoic, they became very numerous and left a dense fossil record that was to persist across the Paleozoic-Mesozoic boundary and last right up through the entire Mesozoic. Then they, too, finally became extinct in the very same set of events that took the dinosaurs away. Geologists love ammonoids, as their dense fossil record and generally rapid evolution allows them to subdivide geological time with fine precision throughout the Upper Paleozoic and, especially, the Mesozoic.

There are three major groups of ammonoids: the goniatites, the ceratites and the ammonites. They differ mostly in the degree of complexity shown by the "suture" lines of the shell. (The sutures are formed by the intersection of internal partitions with the outer shell of the ammonoid). The goniatites appear first and dominate ammonoid faunas until that greatest of all mass extinctions at the end of the Permian Period. Only a few ammonoid species made it across the boundary. [Plate 145] One of them gave rise to the next great ammonoid proliferation—the ceratites, with an advanced, more complex suture design. Then it happened again: ceratites became extinct in the big extinction at the end of the Triassic. Once again, a few species barely made it through, among them one that gave rise to the advanced ammonites, which proliferated and formed a prominent portion of marine life for the rest of the Mesozoic.

PLATE 145

DACTYLIOCERAS TENUICOSTATUM, **AN AMMONOID CEPHALOPOD.**

Lower Jurassic (*ca.* 185 million years). Yorkshire, England.

This species is a relatively early member of the ammonite radiation following the demise of most ceratite ammonoids at the end of the Triassic.

PLATE 146

SYNAPTOPHYLLUM SP., **A COLONIAL RUGOSE CORAL.**

Middle Devonian (*ca.* 385 million years). New York.

Though most Paleozoic rugose corals lived as single individuals, some were colonial—and superficially quite similar to modern colonial scleractinian corals.

Newly evolved players in the ecosystems established after major mass extinctions may be—but often are not—the direct descendants of their recently extinct predecessors. Birds, not mammals, descended from dinosaurs—yet it was the mammals that took over the main terrestrial niches formerly occupied by the familiar dinosaurs after the K-T extinction. In milder extinctions, descendant species may simply replace their own close relatives—as seems to have been the case (as we saw in Chapter 3) when the descendant, seventeen-column Phacops species replaced its presumed ancestor after the seas dried up in the North American Midwest back in Middle Devonian times. There are intermediate cases—where ecological successors may be collateral kin whose adaptations need some overhaul before they can replace their newly extinct relatives. Coral evolution provides a good example of replacement, following mass extinction, by newly evolved kin who then take the ecological place of their relatives who had succumbed in a mass extinction.

For years, it has been known that Paleozoic corals became extinct during the Permo-Triassic crisis. Shortly thereafter, in the Triassic, corals reappear—and, among other things, once again begin the business of building reefs. [Plate 146] For years, experts debated how the Paleozoic rugose corals could have given rise to the Triassic species, which are the direct ancestors of today's modern, or "scleractinian," corals. [Plate 147] The two groups differ radically in internal anatomical design and even have different mineralogical forms of calcium carbonate in their stony skeletal tissues. We now know that the closest relatives of modern corals are actually the naked sea "anemones," which look like corals, except that they lack the hard skeletal tissues of true corals. The betting is that modern corals arose when naked forms evolved a hard, calcareous skeleton, something that occurred only after their ecological counterparts of the Paleozoic, the rugose corals, had become completely extinct. [Plate 148]

PLATE 146

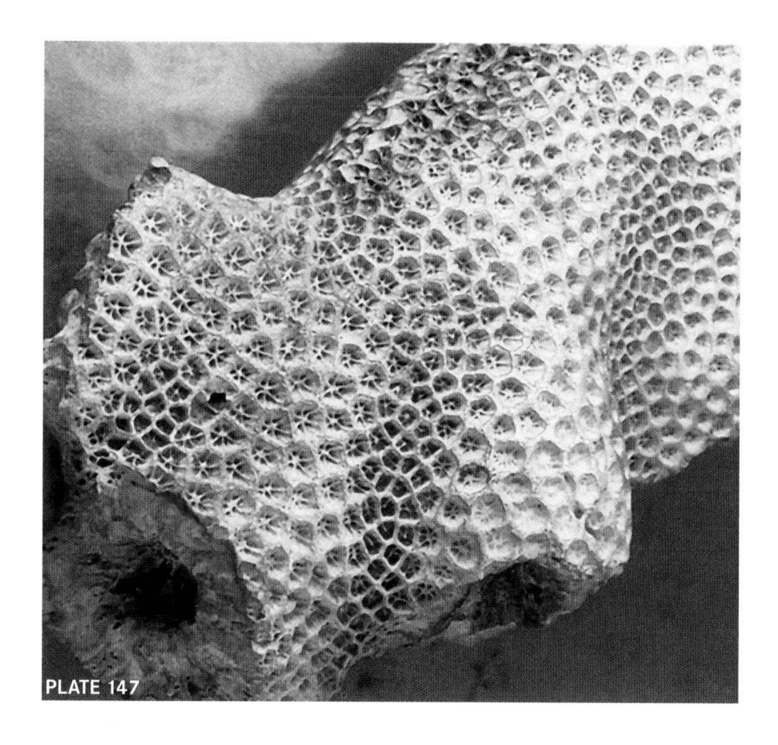

PLATE 147

***SEPTASTREA MARYLANDICA,* A COLONIAL CORAL BELONGING TO THE MODERN GROUP KNOWN AS SCLERACTINIANS.**

Upper Miocene (*ca.* 8 million years). Maryland.

Despite their superficial similarity to Paleozoic tabulate and rugose corals, the Mesozoic and Cenozoic scleractinians are probably most closely related to modern sea anemones, which lack hard skeletal structures.

PLATE 148

***HALYSITES SP.,* A COLONIAL ("TABULATE") CORAL.**

Middle Silurian (*ca.* 425 million years). Clark County, Iowa.

This, the beautiful "ribbon candy" coral, belongs to a group that always lived colonially. Some tabulates formed massive heads reminiscent of brain corals of modern seas.

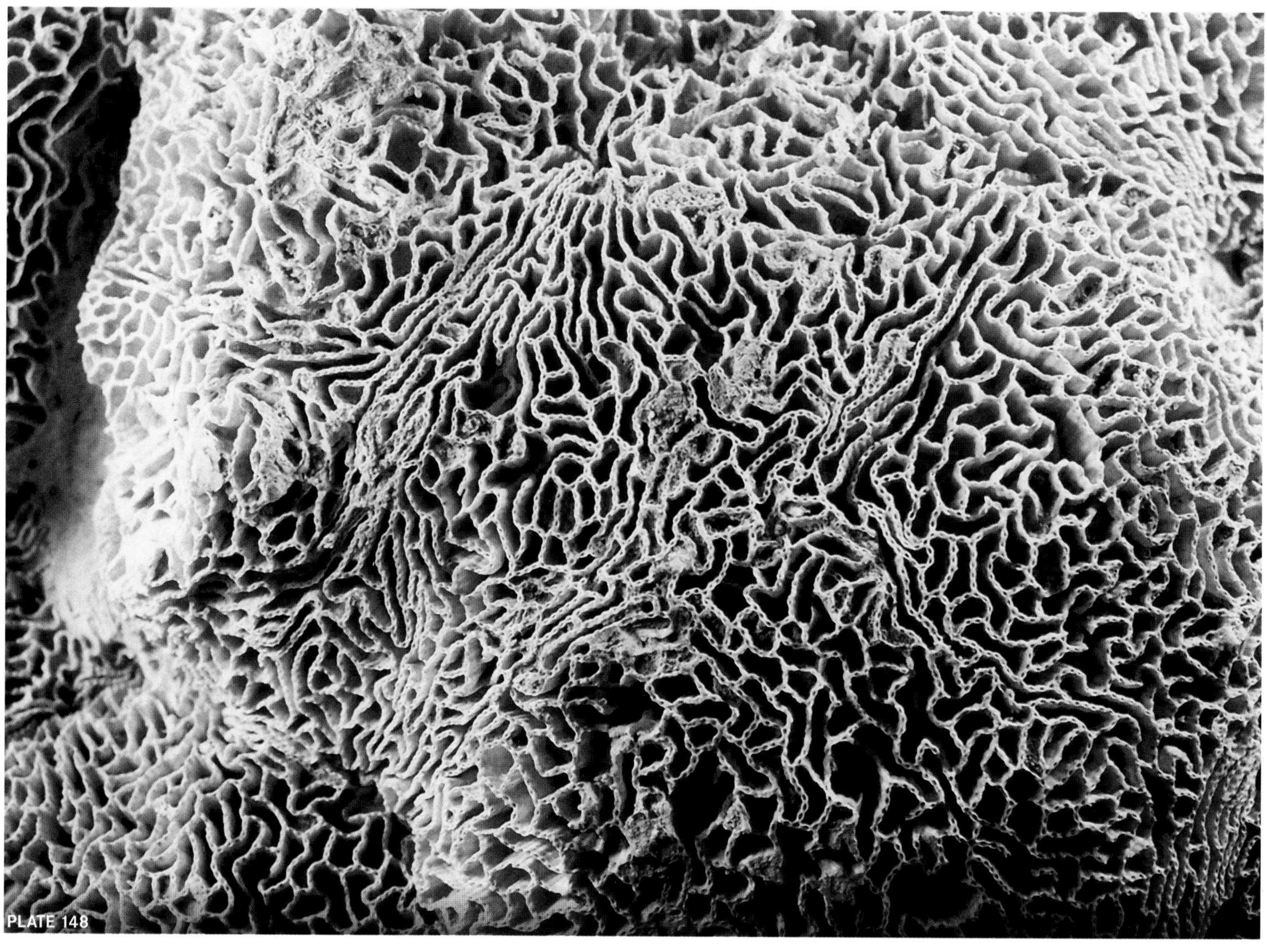

Extinction clearly offers the kind of evolutionary opportunity that we have seen also comes into play when organisms invade entirely new and previously unoccupied ecological realms. Major evolutionary change, whether it is within well-established groups, or involves the origin of groups with novel adaptations, nearly always happens after major extinction events. Nor are the extinctions simply of isolated lineages: the sorts of extinctions that really have a major effect on subsequent evolution are precisely those that cut huge swathes across entire regional, and sometimes even global, ecosystems. In other words, the true mass extinctions of the kind we examined in the preceding chapter.

MACROEVOLUTION: THE SLOSHING BUCKET

Little in the way of significant evolutionary innovation occurs unless extinction has at first cleared a considerably wide ecological path. Think of it: life in ecological communities goes on, generation after generation. Thousands of years go by, and there is, no doubt, some degree of evolutionary change within most of the species that form parts of those ecosystems. Yet the species, as we have seen, tend to remain essentially recognizably the same.

Some species become extinct as the ages roll by: presumably, they could no longer locate recognizable habitat as the inevitable environmental change switches the distribution of habitats around. New species, too, emerge. Those that are ecologically distinct sufficiently to find an ecological foothold manage to survive. These are, for the most part, species that differ, to some perceptible degree, from their ancestral species. This is the level of "punctuated equilibria," with species showing little or no appreciable change during their histories. Occasionally, well-established species give rise to new, reproductively distinct descendant species. Anatomical change, modest as it usually is, seems to be concentrated around speciation events. This effect simply means that only those new reproductive groups—those new species—that are ecologically (hence anatomically) sufficiently distinct from their ancestors will survive and thrive sufficiently well to show up in the fossil record in the first place.

Occasionally (and here we begin to reach aspects of macroevolution) a group of related species within a lineage will seem to take an anatomical "idea" and develop it through time. As we saw in Chapter 4, species can be "sorted" in various ways within evolving lineages, and complex adaptations may develop as the outcome of a series of successive speciations. Recall that brain size increased in human evolution over the past 6 or 7 million years, even though brain size was evidently fairly stable within individual species. What we see is a pattern of better survival rate for larger-brained species within the human lineage through time. What we do *not* see is evidence of a slow, gradual and eminently progressive increase in brain size within a single hominin line advancing steadily from a point in time 6 or 7 million years ago right up to the present day.

But, for the most part, it is business as usual within these long-standing ecosystems as geological time keeps ticking away. As we saw in Chapter 2, paleontologists know that life actually comes in a series of quite distinct "packages" in the succession of marine communities of the Paleozoic. The packages are stacked up, one after another, in geological time. An inland sea, covering the better part of half a continent, will wax and wane, supporting a series of interconnected communities with hundreds of different species. The habitats and communities, as we have seen, do shift around. But the overall structure of the biota—this complex mélange of generally unrelated species—remains pretty much the same for *millions* of years. As we have seen, the phenomenon of stasis describes the typical history of individual species. Because most species show stasis to some degree or another, it will come as no surprise that the overall aspect of the entire ecological system remains quite stable and "recognizably the same."

For millions of years, it is business as usual in ecological terms—and that means in evolutionary terms as well. Stability in evolution begets ecological stability. Or is it the other way around: does long-term ecological stability confer long-term evolutionary stability? Ecology and evolution are interlocked in an intricate web of cause and effect, and there may be no real answer to our question.

But imagine this: suppose that environmental change never reaches a threshold effect. Suppose most species most of the time are perfectly able to find suitable habitat. What then? The fossil record proclaims loudly that it will be business as usual. Those packages of life in the Paleozoic lasted anywhere from 5 to over 10 million years! What ends each package—and simultaneously marks the beginning of the next? *Extinction*: true, cross-genealogical ecological extinction, that's what.

This means that extinction is absolutely vital to the evolutionary process. [Plate 149] Relatively mild extinction events, those that encompass relatively small regions and eliminate relatively few and rather lower-level groups (species and genera, say), are typically followed by the evolution of new species and genera. These newcomers are the staffers of the newly reconstituted ecosystems–and, as a result, the new ecosystems are not altogether that different from the previous, disrupted ecosystems (assuming, that is, that the general nature of the habitats are comparable across the extinction boundary).

Evolutionary change in the long run seems to be linked to extinction; starfish, though they have shown some evolutionary modification during their nearly 500 million year sojourn on Earth, have nonetheless persisted in markedly the same basic form, and performing essentially the same ecological roles, throughout their entire history.

Larger, more inter-regional extinctions, with their great effects on number and size of groups that disappear, have correspondingly greater effects on subsequent evolution. Extinction events that remove entire families and orders, groups ranked rather highly in the Linnaean hierarchy, are followed by such evolutionary proliferation and adaptive innovation that entire new groups of comparable rank are seen to come into existence. The greatest mass extinctions of all occasion the greatest change in the evolutionary scheme–and in the complexion of the ecosystems that follow.

Now think again of the mammals, lucky enough to last long enough that they were still around when the dinosaurs finally yielded control of the Earth's terrestrial ecosystems. After a brief lag, mammals proliferated wildly into a great array of what are now rather dismissively called "archaic" (ancient and outmoded) forms. Most of these failed to survive the Paleocene (the first division of Cenozoic time). But some did survive, and these quickly became our rodents, ungulates, carnivores, elephants–and, yes, our whales and bats. [Plate 150] Nothing nearly so significant in mammalian evolution has subsequently happened in the 50-odd million years since the Eocene. [Plate 151]

PLATE 149

SGONIASTER SP.,
A STARFISH ECHINODERM.

Upper Cretaceous (*ca.* 80 million years). California.

Evolutionary change in the long run seems to be linked to extinction; starfish, though they have shown some evolutionary modification during their nearly 500 million year sojourn on Earth, have nonetheless persisted in markedly the same basic form, and performing essentially the same ecological roles, throughout their entire history.

PLATE 149

PLATE 150

PLATE 150

***CORYPHODON* SP.,**
AN ARCHAIC MAMMAL.

Lower Eocene (*ca.* 50 million years).
Big Horn Basin, Wyoming.

This was one of the early "archaic" mammalian forms that in turn became extinct, to be followed at last by the evolution of modern groups of mammals.

PLATE 151

METACHEIRONOMYS TATUSIA,
A PRIMITIVE MAMMAL RELATED TO ARMADILLOS.

Middle Eocene (*ca.* 45 million years). Wyoming.

This species belongs to the edentates ("toothless ones"), among the most primitive of all placental (or "true") mammals.

PLATE 151

Save ourselves, of course. We, the clever ape, really do represent some degree of novelty. And even in our own evolution, as we have already seen, there is clear evidence that extinction played a major role. How long we manage to stick around will determine how much in retrospect (to who- or whatever is looking at the fossil record in the future) we ourselves constitute a legitimate case of *macroevolution*.

I call this pattern the "Sloshing Bucket" theory of evolution. Evolution is contingent on environmental change. Small perturbations—like fires, or volcanoes, may be divesting: Krakatoa evidently killed all plants, animals and microbes living on its flanks when it exploded. But when things cooled down, recruitment of the same species from neighboring areas rebuilds ecosystems more or less identical to the one that was destroyed.

But when ecological disturbance is sufficiently intense and more importantly, widespread, entire species are in peril. Extinction can claim large numbers of species in regional ecosystems, more or less simultaneously. This is what paleontologist Elisabeth Vrba called "turnover pulses," as recovery in part depends on migration in of surviving species. But these spasms of regional species extinctions trigger episodes of speciation: evolution begins to play a role in the rebuilding of new ecosystems. These events, in my view, are the main locus of adaptive change in the evolution of life. [Plate 120, see page 192; also Plates 132, 133. 134]

And, finally, there are the truly global mass extinctions. They are few and far between—but their evolutionary effects are of a magnitude far greater than the regional turnovers. Hence the "Sloshing Bucket"—the greater the scope and scale of ecological disturbance, the greater the evolutionary response. The greater the evolutionary response, the greater the differences in the nature and composition of the ecosystems once the world settles back down to the stability of "normal" times. It is a picture that reminds me of the degree that water will slosh from side-to-side when the bucket is disturbed.

EPILOGUE
Old and New Pictures

We have neared the end of a pictorial double album of ancient life. The photographs in *Extinction and Evolution* have been arranged to complement the word pictures: patterns in life's history, and what they imply for our understanding of the nature of evolution. The ideas flesh out the bare bones and dry shells—just as the fossils themselves are the primary documents from which all these ideas spring. Paleontology, like any science, is an interplay between material reality and ideas. [Plate 152]

Indeed, the very perception of that reality—simply looking at a fossil—immediately engages the mind in interpretation. *Everything* that we think we know about the material universe really is, at base, an idea. All but a few flat-earthers know that Earth is spherical, rotates on its axis and revolves around the sun. Yet, not too many years ago, such a picture of the physical conformation and behavior of Earth was considered dangerously heretical and patently false. That it is now generally considered as "fact" does not alter its status as an "idea."

Evolution, of course, is an idea, a picture that explains why there is a pattern of similarity interlinking all forms of life, living and extinct. It is the only scientific explanation (barring, that is, the supernatural) for the sequence of life revealed in the fossil record, and for all the experimental work in agricultural, medical and genetics laboratories showing that heritable traits of organism can and do change under known conditions. We call evolution a "fact," but we should never forget that, like anything else we say about nature, evolution is an idea—a picture. A fact is nothing but an idea that is so well established that it is no longer in doubt. The evolution of life, along with the idea that Earth is incredibly old and itself has had a history, are the two fundamental pictures of this book. Together, they constitute nothing short of two of the greatest ideas in the brief history of human thought.

But we have also seen that some, at least, of the conceptual pictures I have drawn in these pages are by no means so universally accepted as the grand ideas of evolution and the hoary age of Earth. I have given here my own perception on the basic patterns of the fossil record, and my thoughts—as they have matured so far—on what those pictures mean for a better understanding of the very nature of the evolutionary process. Some of these ideas are more generally accepted than others. Naturally enough, the closer to the empirical pictures—the patterns of stability and change, of appearance and disappearance—the less controversial are these ideas.

In recent years, the two most important contributions to modern evolutionary thought coming from paleontology each have strong empirical components. The first is stasis—the notion that, once they appear, species do not as a rule tend to accumulate a great deal of subsequent evolutionary change. It is an important ingredient of the more general idea that evolutionary change tends to be concentrated in relatively abrupt intervals. [Plate 153] As we saw in Chapter 3, stasis was not anticipated by evolutionary geneticists—nor, at first glance, does it seem to agree with the expectation of inevitable slow, steady progressive evolutionary change as the very quintessence of the original Darwinian picture of evolution. Stasis, recall, was fully recognized by nineteenth century paleontologists—but its importance was overlooked and the phenomenon as a whole was disregarded, presumably because it conflicted so sharply with the Darwinian picture. Recall, too, that the pattern of stasis coupled with brief spurts of (generally modest amounts of) evolutionary change—the picture of "punctuated equilibria"—is conceptually related to George Simpson's picture of large-scale evolutionary change likewise coming in relatively rapid spurts, when compared with the much longer periods of time of more sedate rates of evolutionary change within large-scale groups of organisms: Simpson's notion of "quantum evolution."

PLATE 152

***ENCHODUS* SP., A LIZARD FISH.**

Upper Cretaceous (*ca*. 80 million years). Green River Formation, Colorado

The genus *Enchodus* survived into the early Tertiary, making it through the mass extinction event that eliminated terrestrial dinosaurs, ammonites and many other groups. Clearly fierce predators themselves, species of *Enchodus* were themselves preyed upon by larger animals, including elasmosaurs. Thus they appear to have played a major role in the food chain in Cretaceous seaways.

Credit: Roger Weller, Cochise College.

Both Simpson and later paleontologists (such as Stephen Jay Gould, Steven M. Stanley, Elisabeth S. Vrba, Bruce Lieberman and myself) thought that such patterns seen in the fossil record, rather than simply reflecting the incompleteness of this record, actually have something to tell us directly about the nature of the evolutionary process. And that "something" transcends the obvious conclusion that evolutionary rates are variable. Thus Simpson thought a new element of evolutionary theory–his "quantum evolution," involving a novel picture of interactions among evolutionary processes–was called for by the data of the fossil record. Though his interpretations were criticized by prominent geneticists, and though Simpson himself later modified his views, it is interesting that evolutionary geneticists have recently begun to take Simpson's earlier visualizations of quantum evolution very seriously.

So too with punctuated equilibria. Geneticists have recently been adjusting their models, their mathematical pictures, of the evolutionary process to embrace this empirical phenomenon. And recall that, unlike Simpson and his picture of quantum evolution, Steve Gould and I merely adopted another aspect of conventional evolutionary theory–the notion of "speciation"–to explain the relatively rapid bursts of evolutionary change that punctuates the more protracted periods when little or no change occurs. That was (and remains) our interpretation, our explanatory picture of the pattern. It must be said that, while biologists of course universally recognize the phenomenon of speciation, some remain reluctant to accept that evolutionary change in general is concentrated heavily within episodes of speciation. I think the fossil record proclaims this to be true, but I also understand the difficulty in seeing how and why adaptive change in general should be focused mostly around episodes where new reproductive communities–new species–are being formed from ancestral species. Here we come to a current frontier in theoretical evolutionary biology, one on which I and others have been concentrating recently.

The second great idea with a strong empirical component to come from paleontology in recent years is also a resurrection of a pattern well understood, even in pre-Darwinian days, in the nineteenth century, but one which was somehow left to molder in obscurity for well over a century: *mass extinction*. As we have seen, there can be no question that mass extinctions have, from time to time, and with varying degrees of severity, disrupted the biota. [Plate 154] The effects invariably cut right across genealogical lines of descent, and there can be no doubt that extinction is an inherently ecological phenomenon. From an evolutionary perspective, however, we can now see vividly how much evolution is dependent upon mass extinction. Little evolution is likely to occur unless extinction has greatly shaken up–and diminished the numbers of–species already in existence. This picture contrasts mightily with the traditional image of the history of life progressing more or less smoothly, with new groups appearing regularly, and adaptive change accumulating in a stately and wholly regularized fashion.

These two grand pictures–evolutionary change restricted as a rule to relatively brief episodes, and mass extinction–are obviously closely interrelated. Together they rather radically alter our views of the basic nature of life's history. They invite us (I would prefer to say, actually *force* us) to tinker with the conceptual picture of how the evolutionary process actually works. And this is an exciting and very much ongoing process.

For example, consider the question: Why do speciation rates increase after mass extinction–as they must do when life reproliferates after such extinction episodes? For much of my career, I shared the common explanation to this question. I thought that extinction frees up ecological niches, into which new species radiate. But I could not understand *how* so-called "empty niches" could actually trigger speciation–which, recall, entails the establishment of new reproductive communities. The problem has been bothering me for years, and I had made only some incomplete progress towards its solution.

The answer was achieved by Yale paleontologist Elisabeth Vrba, who was able to sidestep the old vision of "speciation into vacant niches" and look at the problem in an entirely different–and much simpler and comprehensible–way. Instead of seeing speciation driven in some fashion by a "need" to fill empty niches, she realized that the very same factor that causes extinction promotes speciation: the alteration of the distribution and nature of habitats. Climate change lies at the heart of nearly all extinction events. Climate change is the cause *par excellence* of the alteration and distribution of habitats. Species become extinct when accustomed habitats can no longer be found. And habitat fragmentation is the *sine qua non* of species fragmentation: the essence of speciation. [Plate 155]

PLATE 153

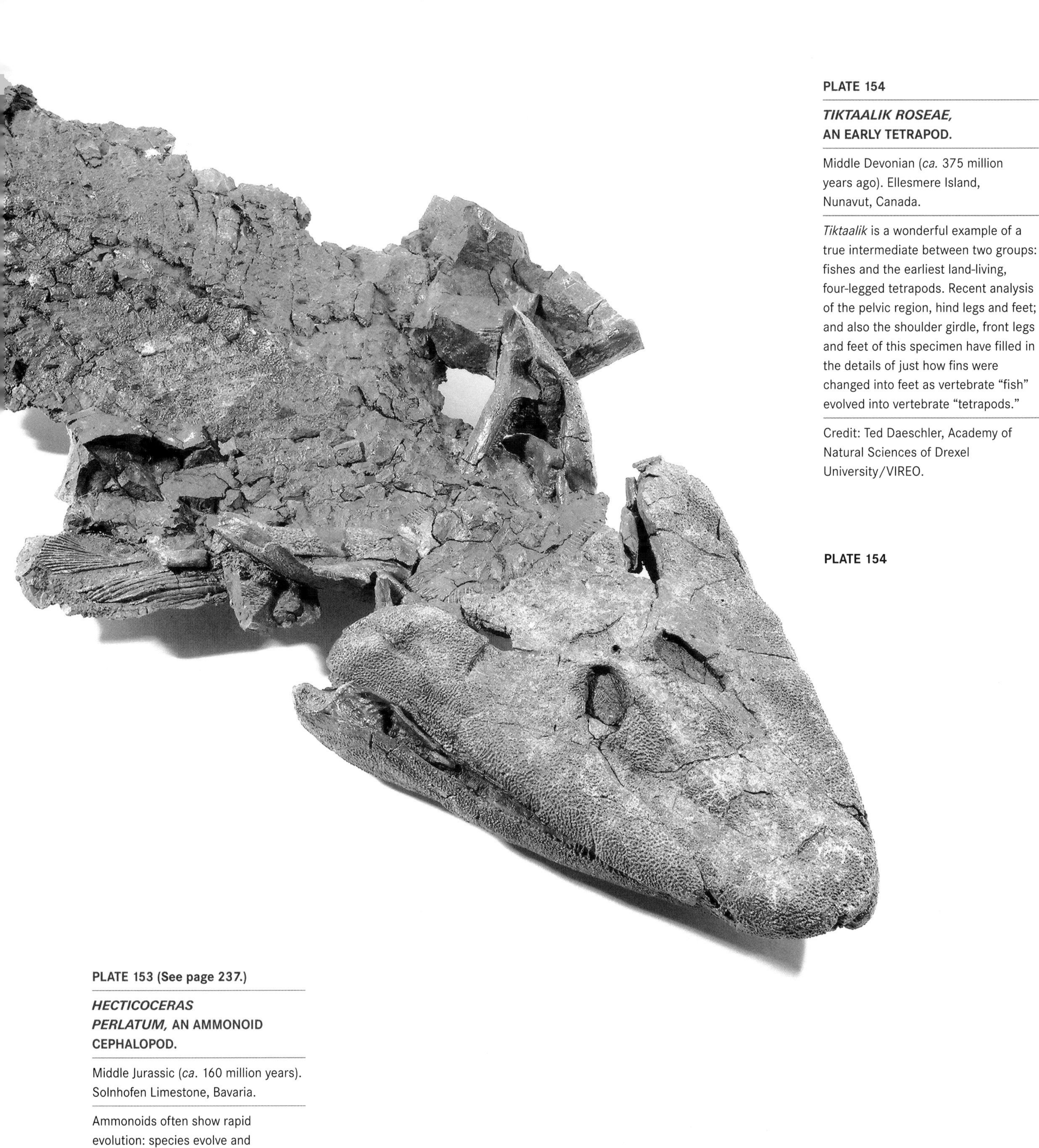

PLATE 154

***TIKTAALIK ROSEAE,* AN EARLY TETRAPOD.**

Middle Devonian (*ca.* 375 million years ago). Ellesmere Island, Nunavut, Canada.

Tiktaalik is a wonderful example of a true intermediate between two groups: fishes and the earliest land-living, four-legged tetrapods. Recent analysis of the pelvic region, hind legs and feet; and also the shoulder girdle, front legs and feet of this specimen have filled in the details of just how fins were changed into feet as vertebrate "fish" evolved into vertebrate "tetrapods."

Credit: Ted Daeschler, Academy of Natural Sciences of Drexel University/VIREO.

PLATE 154

PLATE 153 (See page 237.)

***HECTICOCERAS PERLATUM,* AN AMMONOID CEPHALOPOD.**

Middle Jurassic (*ca.* 160 million years). Solnhofen Limestone, Bavaria.

Ammonoids often show rapid evolution: species evolve and become extinct at faster rates than many other groups, such as the trilobites of the Paleozoic.

Nothing could be simpler. There is great appeal to an idea that sees a single, if complex, causal agent accounting for both sides of the pattern: extinction *and speciation*. Extinction and speciation are now more intimately linked than ever before, thanks to the theoretical insights of Elisabeth Vrba.

And so the beat goes on. Renewed appreciation of the big pictures of the fossil record continue to send mild shock waves against the canons of traditional thought. It is a stimulating challenge—and downright fun to boot. I and like-minded colleagues will continue to adjust the old picture until it comes more sharply into focus with the basic patterns we see in the fossil record. Large-scale switches in the perception of pattern create open vistas for rethinking old ideas: the conceptual equivalent of hitting on a vein of high grade ore. The conceptual mine portrayed in these pages is far from completely quarried out!

Yet there is a flip-side to the excitement of discovery afforded by a fundamental shift in looking at nature. It must always be conceded that these ideas are pictures too—and subject to the same revision, or perhaps even abandonment, at some future date. When Steve Gould and I published our original paper on punctuated equilibria in 1972 (a follow-up, it must be said, of my original very similar paper in 1971), we acknowledged our frank attempt to substitute one picture (stasis plus brief spurts of change) for another—the Darwinian expectation of gradual, progressive change. In our second paper, in 1977, we speculated that Darwin's vision was colored by the Zeitgeist of his times: the Industrial Revolution was well underway, and a belief in progress was most definitely in the air. It had become acceptable to imagine change in systems (social as well as organic), particularly if change is orderly, gradual, and represents a form of palpable improvement.

But, we acknowledged, if such be arguably true of Darwin, it must be true of all of us when we confront nature—and to demonstrate that we certainly did not think we were exempt from the effects of ideas more generally held in society at large impinging on our own perceptions of nature, we wrote that "it may also not be irrelevant...that one of us learned his Marxism, literally at his daddy's knee." Punctuated equilibria, after all, sees change coming in brief spurts, following long fallow periods of the status quo—and thus bears a superficial similarity to Marx's vision of human socioeconomic historical pattern.

Reaction to this comment was quick in coming. Punctuated equilibria was seen in some circles as pure Marxist propaganda—and rejected in others as fake, misconstrued Marxism! It was neither, of course: Darwin's was an honest attempt to paint a general picture of the history of life. The source of that picture may not have been wholly the fossils themselves, nor even the straightforward extrapolation of his experiences as an apprentice pigeon fancier. There may also have been those additional *Zeitgeist* elements of progress—though it is my belief that such would enter in, not as a source for the imagery, but as a welcome parallel that may have allowed Darwin to feel more comfortable about springing such a shocking idea as evolution on his world. Nonetheless, his was an empirical claim: he said it himself, that future paleontological research would settle the matter of the basic patterns of evolutionary change. (He actually said that his theory would stand or fall on what future paleontologists had to say on the matter—which was going far too far!). Darwin's picture, whatever its source, was empirical, subject to further testing.

Such, of course, is true of the patterns of stasis and change, origination and extinction, that I have dwelled upon here. It was not Marxism, I believe, but the emergence of parallel views in the 1960s and 1970s in unrelated disciplines that may be analogous to the similarity that Darwin's thoughts had to more general notions in the air in his time. In the 1970s, mathematics had its "catastrophe" theory, for example; and there were parallel developments in economics. Closer to home, catastrophism in geology was receiving a new look—for the first time, really, since Lyell's *The Principles of Geology* were published in the 1830s. Most telling of all was the publication, in 1962, of Thomas Kuhn's immensely influential *The Structure of Scientific Revolutions*. Kuhn sees progress in science as restricted, for the most part, to revolutionary "paradigm shifts"—wholesale substitution of one picture of nature for another. These revolutions, moreover, are envisioned as coming pretty abruptly, following long periods of little or no real change in scientific outlook. Stasis punctuated by brief spurts of change became a common picture in many different academic disciplines just about the time we were formulating our own views. If there is a parallel to be found, this is it—not Marx. Steve and I even quoted Kuhn in the original paper.

There is a further connection between Kuhn's view of the nature of scientific work and the emergence of punctuated equilibria. It has become notorious that, no matter what the fate of Kuhn's ideas among historians of science (less enthusiasm now than when initially promulgated, I believe—time does march on!), no scientist since Kuhn has ever wanted to be anything less than a revolutionary. And here we were, in 1972, openly calling for a picture change (though we were not

PLATE 155

PSITTACOSAURUS MONGOLIENSIS,
AN HERBIVOROUS DINOSAUR.

Cretaceous (*ca.* 100 million years). Mongolia.

Both extinction and speciation follow habitat fragmentation and alteration—no doubt as true for this extinct species as for any other (See also Plate 30).

so bold or egotistical to call it a "Kuhnian paradigm shift"). To this day, critics still write that punctuated equilibria may well find a place in the annals of evolutionary biology—but that it is by no means the grand sort of shift of vision that its proponents claim it to be.

Yet my own view of the history of the notion of punctuated equilibria has little to do with parallels in other disciplines, or general cultural phenomena—and still less with notions of revolution in science. Ironically enough, my picture of change in scientific views is much more akin to the old notion of "growth in knowledge": I do believe in a form of progress in scientific knowledge, even though throughout much of my career I have actively opposed the old notion of progress in evolution! Specifically, I think that there is a reality to nature, and that, with tinkering, we adjust both our pictures, and our explanations, of nature. We'll never get completely there, but we are indeed truth seekers, and if we did not think that we were getting somewhere, collectively, with our attempts to depict and understand nature, then there would be little point in pursuing science. I do not believe that one picture of nature is as good as another. In that spirit, I see punctuated equilibria as the amalgamation and extension of the approaches and some of the insights of George Simpson, melded with the notion of species and speciation developed by his contemporaries, Ernst Mayr and Theodosius Dobzhansky. We took Simpson's zest for paleontological pattern, saw that it held, critically, to be true within species, and then embraced (as Simpson never did) the Dobzhansky-Mayr vision of species as reproductive communities as a means of explaining the stasis-and-spurts-of-change pattern we hold to be typical of species in the fossil record. Punctuated equilibria, no more than any idea, did not come' out of thin air—and one need not look far to identify its intellectual roots.

PLATE 156

CRANIOCERAS SKINNERI,
A TERTIARY ARTIODACTYL (CLOVEN-HOOFED) MAMMAL.

Miocene (*ca.* 12 million years). Crockett, Texas.

The fossil record continues to surprise, delight and inform us with its richness of diversity, its puzzles—and its clues to the history of life, and to the processes that underlie that history.

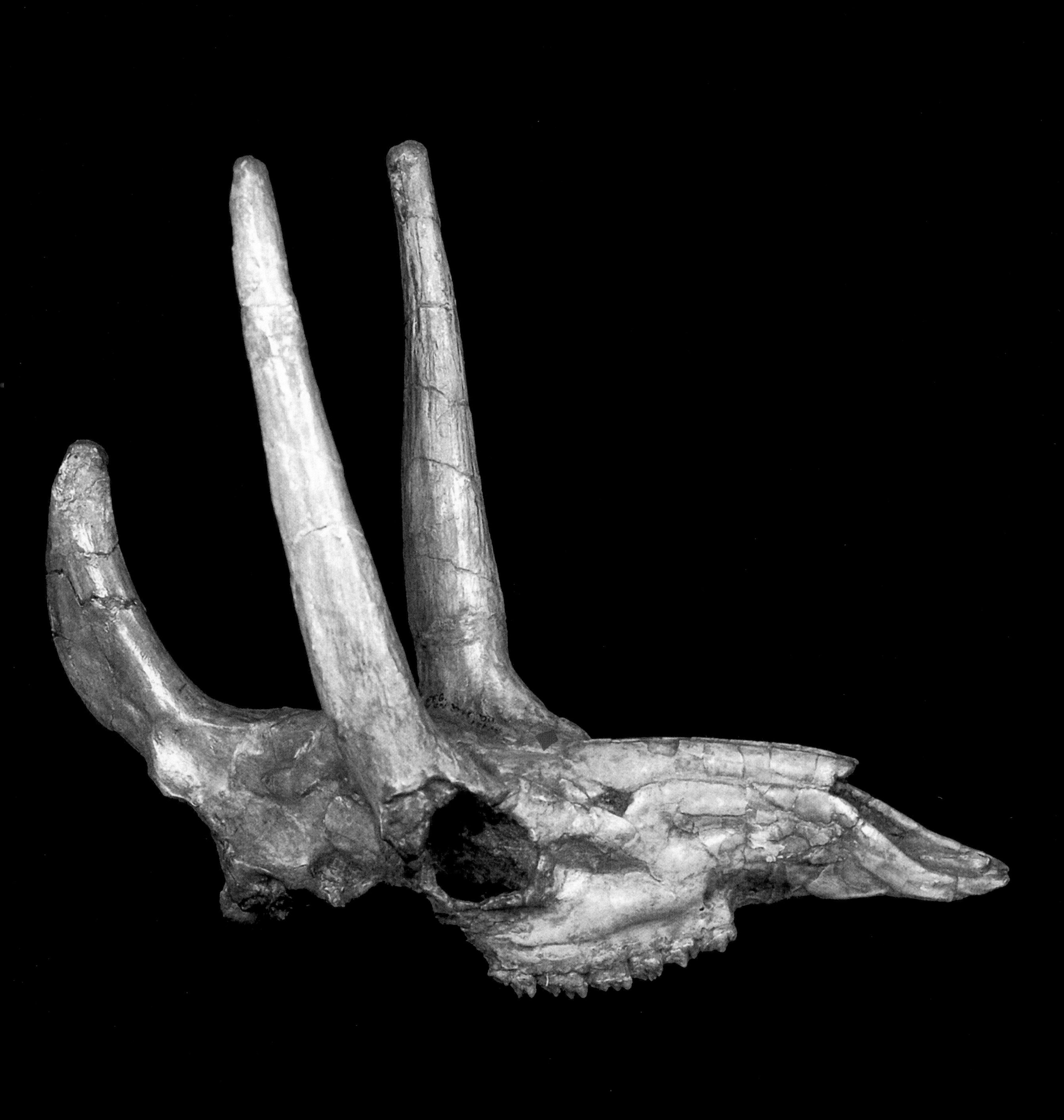

PLATE 156

PLATE 157

HELIOBATIS RADIANS,
A STINGRAY.

Lower Eocene (*ca.* 52 million years). Fossil Lake, Green River Formation, Wyoming.

This gorgeous stingray fossil makes it strikingly clear that stingrays have not changed all that much for at least the past 50 million years—if not even much longer.

Credit: National Park Service.

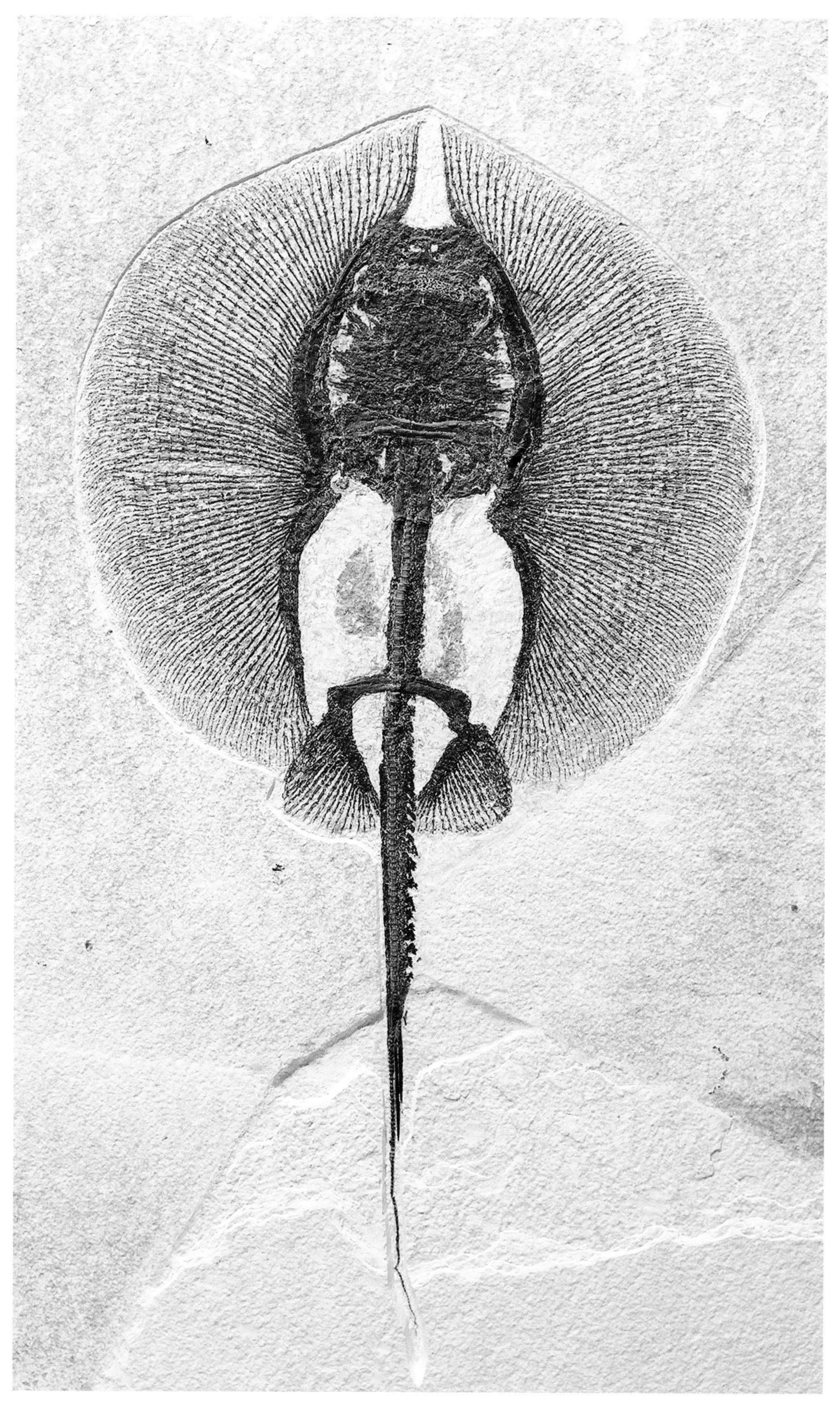

How radically the picture has changed from the days of Dobzhansky, Mayr and Simpson depends on how far along the path one travels. There have been changes made in basic perception of patterns in the history of life. [Plate 156] These appear to me to require some basic rethinking of evolutionary theory. The outcome, though, is very much a matter for the future to decide.

I will continue to mine what I take to be the mother lode of opportunity for (dare I say it?) progress in evolutionary theory. Lately, my work has tended to become very abstract—far more so than the elements revealed in these pages. I have been grappling with the dual organization of life into two hierarchical systems. On one hand, organisms make a living and interact with other kinds of creatures, and with the physical world, placing them automatically in an ecological world: local populations of species interact with populations of other species to form local ecosystems; local ecosystems interact to form larger regional ecosystems. [Plate 157]

On the other hand, that organisms reproduce means that they are parts of local reproductive populations—which are networked together into the units we call "species." Species give rise to descendant species, forming chains of ancestry and descent that we call "higher taxa"—genera, families, orders and phyla. The latter sort of genealogical system is the traditional focus of evolutionary biology—while ecologists look at the intricate networks of ecological interactions cutting across genealogical lines. There has been confusion on how these two different sorts of systems interrelate, yet it is interaction between elements of both that has given us the history of life.

But no scientist can safely get too far from the empirical world, where mental pictures are tied rather more concretely to patterns inherent in nature. I am going back to the fossil record, these days with an ecological slant, to look at the relation between stability and change that I know is typical of most species, and patterns of stability and change that may hold true for entire ecological communities. Therein should lie some solid clues as to just how the ecological and evolutionary realms are related. The beat indeed goes on.

PHOTOGRAPHER'S STATEMENT
For Lily: Timeless Love

Years ago, while trekking on the Himalayan pass of Thorong La in Nepal, my Sherpa guide took great delight in picking up stones and smashing them in half with larger stones to show me fossils contained within. I remember thinking: what are these strange sea creatures doing at 17,800 feet encased in a slab of rock? They appeared to be frozen in time, not unlike the early dwellers of Pompeii.

In approaching this project it's important that the photographer retain that sense of surprise and mystery. Looking at fossils is like peering through a window into the past. They should be photographed in a way which arouses the curiosity and wonderment of the viewer with that element of surprise not unlike the one I experienced that day on the high pass in Nepal.

While photographing ancient skulls and bones at the American Museum of Natural History, my thoughts drifted to the words of Hamlet: "That skull had a tongue in it and could sing once..."

Constantly reminding myself that these fossils were once living breathing beings, I was overcome by a sense of a profound mystery working out its destiny. I felt part of the majestic journey that life has taken here on Earth.

Murray Alcosser
New York, New York

Murray Alcosser died in 1992. Born in Brooklyn, New York in 1937, he began his working life as an electrician. In the mid 60s, he was one of the supervisors who oversaw the wiring of the new Madison Square Garden. All along, however, he thought there was a photographer inside of him and ten years later left his lucrative career as a master electrician. His creative talents were soon recognized with his photographs featured in books published by Rizzoli International and Harry N. Abrams and in major exhibitions. Although we have replaced a number of his original photographs from the first edition of this book with those of recently discovered fossils, Alcosser remains one of the preeminent natural history photographers of our times. He is missed by all who knew him.

PLATE 158

UROGOMPHUS GIGANTEUS,
A DRAGONFLY.

Upper Jurassic (*ca*. 150 million years). Solnhofen Limestone, Bavaria.

The esthetic aspect of fossils is by no means the least of their value to us.

PLATE 158

PLATE 159

MACGINITIEA WYOMINGENSIS,
A SYCAMORE TREE LEAF.

Midddle Eocene (*ca*. 50 million years). Green River Formation, Douglas Pass, Colorado.

Credit: Steve Wagner, Paleocurrents.

"All religions, arts and sciences are branches of the same tree. All these aspirations are directed toward ennobling man's life, lifting it from the sphere of mere physical existence and leading the individual towards freedom."

– Albert Einstein,
Out of My Later Years

ANNOTATED BIBLIOGRAPHY

Carroll, Sean. 2009. Remarkable Creatures: Epic Adventures in the Search for the *Origins of Species*. Mariner Books, Boston.

Carroll provides vignettes of some of the fascinating people who have made the most significant contributions to understanding the evolutionary processes: Alexander von Humboldt, Charles Darwin, Alfred Russel Wallace and Henry Walter Bates, Eugène Dubois, Charles Walcott and Neil Shubin, Louis and Mary Leakey, Linus Pauling and Allan Wilson.

Dawkins, Richard. 1986. *The Blind Watchmaker: Why the Evidence of Evolution Reveals a Universe without Design*. W. W. Norton & Co, New York.

An ultra-Darwinian refutation to the argument-by-design, most notably made by the 18th century theologian William Paley, that the universe, like a watch in its complexity, needed, in effect, a watchmaker to design it.

DeSalle, Rob, and Ian Tattersall. 2008. *Human Origins: What Bones and Genomes Tell Us about Ourselves*. Texas A&M University Press, College Station.

Illustrates how traditional paleontology and modern genetics tell the same story of human evolution.

Dobzhansky, Theodosius. 1937. *Genetics and The Origin of Species*. Columbia University Press, New York (reprint edition, 1982).

One of the cornerstone volumes of the "synthetic theory of evolution," integrating genetics with the Darwinian vision, and exploring the biological nature of species.

Eldredge, Niles. 1971. "The allopatric model and phylogeny of Paleozoic invertebrates." *Evolution* 25: 156-67.

The original paper which challenged Darwin's gradualistic view of evolution.

__________, and Stephen J. Gould. 1972. "Punctuated equilibria: an alternative to phylectic gradualism," in T.J.M. Schoft (ed.), *Models in Paleobiology*, pp. 82-115. Freeman, Cooper and Co., San Francisco.

A follow-up to Eldredge's 1971 episodic model, for the first time named Punctuated Equilibria.

__________. 1985. *Time Frames: The Rethinking of Darwinian Evolution and the Theory of Punctuated Equilibria*. Simon and Schuster, New York.

An engaging account of the research that led to recasting the tempo and mode of Darwinian Natural Selection.

__________. 1985. *Unfinished Synthesis: Biological Hierarchies and Modern Evolutionary Thought*. Oxford University Press, New York.

Provides a critical analysis of the works of Dobzhansky, Mayr and Simpson (cited here), and develops a hierarchical view of biological systems and a matching, extended theory of evolution that expands on the insights of the evolutionary synthesis.

__________. 2014. *Eternal Ephemera: Adaptation and the Origin of Species since 1801*. Columbia University Press, New York.

The history of the development of the evolutionary ideas presented in this book, including the works of early paleontologists Lamarck and Brocchi; Charles Darwin; and the origin and development of "Punctuated Equilibria" and the "Sloshing Bucket" theory of evolution.

Futuyma, Douglas J, 1986. *Evolutionary Biology*. Second ed. Sinauer Associates, Sunderland, Massachusetts.

An outstanding college-level text on evolutionary biology.

Johanson, Donald, and Blake Edgar. 2006. *From Lucy to Language*, 2nd edition. Simon and Schuster, New York.

A family album of our human ancestors from 7 million years ago to the present. Features luminous photographs the most significant human fossil skulls all reproduced at life-size.

Mayr, Ernst. 1942. *Systematics and the Origin of Species* Columbia University Press, New York (reprint edition, 1982).

The second founding document of the evolutionary synthesis, focusing on species and their evolution.

Redmond, Ian, 2011. *The Primate Family Tree: The Amazing Diversity of Our Closest Relatives*. Firefly Books, Richmond Hill, Ontario.

A comprehensive resource on the subject of our animal relatives: apes, monkeys and lemurs.

Shubin, Neil. 2008. *Your Inner Fish: A Journey into the 3.5-Billion-Year History of the Human Body*. Pantheon, New York.

Neil Shubin's research expeditions around the world have redefined the way we now look at the origins of mammals, frogs, crocodiles, tetrapods, and sarcopterygian fish—and thus the way we look at the descent of humankind.

Simpson, George Gaylord. 1944. *Tempo and Mode in Evolution*. Columbia University Press, New York (reprint edition, 1984).

Evolution from a paleontologist's perspective, the third fundamental document of the "modern synthesis," with the original formulation of Simpson's theory of "Quantum Evolution."

Sepkoski, David. 2012. *Rereading the Fossil Record: The Growth of Paleobiology as an Evolutionary Discipline*. University of Chicago Press, Chicago.

The definitive account of the genesis of the theory of Puncuated Equilibria.

Tattersall, Ian. 2008. *The Fossil Trail: How We know What We Think We Know about Human Evolution*. 2nd Edition. Oxford University Press.

Paleoanthropologist Ian Tattersall's masterful analysis of the entire scope and sweep of hominid paleontology.

__________. 2013. *Masters of the Planet: The Search for our Human Origins*. Macmillan Science, New York.

Tattersall's penetrating analysis of the fossil record of human evolution leading up to the emergence of our own species, *Homo sapiens*.

Vrba, Elisabeth S. 1991. "From ecology to macroevolution: The Red Queen, Turnover-Pulses, and related topics." In Stenseth, N. C. (ed.), *Coevolution in Ecosystems and the Red Queen Hypothesis.* Cambridge University Press, Cambridge.

Paleobiologist E. S. Vrba has contributed significantly to the improvement of fit between patterns in the history of life and our ideas on how the evolutionary process actually works—her "effect hypothesis" and the notion of "turnover pulses" being especially significant. Most of her work thus far is published as technical articles in scientific journals and books; this paper provides an entree into her bibliography.

Zimmer, Carl. 1998. *At the Water's Edge: Fish with Fingers, Whales with Legs, and How Life Came Ashore but Then Went Back to Sea*. The Free Press, New York.

How engines of macroevolution have transformed body shapes across millions of years.

__________. 2001. Evolution: *The Triumph of an Idea*. HarperCollins, New York.

An up-to-date view of evolution that explores the far-reaching implications of Darwin's theory and emphasizes the power, significance, and relevance of evolution to our lives today.

STRICTLY FOR FOSSIL LOVERS

Boardman, Richard S., Alan H. Cheetham and Albert J. Rowell (eds.). 1987. *Fossil Invertebrates*. Blackwell Scientific Publications, Palo Alto, California.

A comprehensive, multi-authored survey of the anatomy and geologic occurrence of invertebrate fossils.

Carroll, Robert L. 1988. *Vertebrate Paleontology and Evolution*. W.H. Freeman and Co., New York.

The up-to-date, authoritative reference on world-wide vertebrate fossil remains.

Clarkson, Euan N. K. 1986. *Invertebrate Palaeontology and Evolution*. Second edition. Allen & Unwin, London.

An excellent and comprehensive text on invertebrate paleontology, world-wide in scope.

Fenton, Carroll Lane, and Mildred Adams Fenton. 1989. *The Fossil Book: A Record of Prehistoric Life*. Second ed. Rich, Patricia V., et al., eds. Doubleday and Company, New York.

An entertaining and easily-read, well illustrated general book covering all facets of the fossil record.

Klein, Richard G. 1989. *The Human Career. Human Biological and Cultural Origins*. University of Chicago Press, Chicago.

A tour-de-force review of the human archeological and paleontological record.

Maisey, John. 1966. *Discovering Fossil Fishes*. Henry Holt and Company, New York.

Traces the evolution of fishes over the course of 450 million years.

Moore, R. C. (ed.). 1953 et seq. *Treatise on Invertebrate Paleontology*. Geological Society of America and University of Kansas Press, Lawrence.

A multi-volume, multi-authored synopsis and illustration of all known genera of invertebrate fossils. The most authoritative such compendium in the world.

Murray, John W. (ed.). 1985. *Atlas of Invertebrate Macrofossils*. John Wiley and Sons, New York.

Black-and-white photographs, systematically arranged, of many of the world's most important invertebrate fossils.

CLASSICS

Darwin, Charles. 1985. *On the Origin of Species by Means of Natural Selection, or the Preservation of Favoured Races in the Struggle for Life*. Penguin Classics, New York.

Reprint of the 1859 first edition. One of the most influential books in the history of science. Its enduring success lies not only in Darwin's impeccable argument but also in its eloquent style in the tradition of 19th century English literature.

__________. 2003. *The Origin of Species*. Signet Classics, New York.

Reprint of the 6th and final edition of this classic in science. Originally published in 1872: revised and retitled by Darwin as it is know to today, *The Origin of Species*.

__________. 2004. *The Descent of Man: Selection in Relation to Sex*. Penguin Classics, New York.

Reprint of the 1871 original edition. Believing the subject too "surrounded with prejudices" to address in *On the Origin of Species*, Darwin waited twelve years to include human evolution in his theory of Natural Selection and put apes in our family tree.

Haeckel, Ernst. 2010. *Art Forms in Nature*. Prestel Publishing, New York.

A facsimile edition of *Kunstformen der Natur* first published between 1899 and 1904, consisting of 100 engraving in color and various hued shades of black and white of a multitude of strangely beautiful natural forms from Radiolaria to hummingbirds.

Hutton, James. 2010. *Theory of the Earth, with proofs and illustrations*, 2 vols. Gale ECCO, Independence, Kentucky.

A facsimile edition of this 1788 book, which marked the beginning of modern geology, wherein Hutton put forward the view that "from what has actually been, we have data for concluding with regard to that which is to happen thereafter," the key concept to what was to become uniformitarianism that was to so influence Darwin.

Lyell, Charles. 1990. *Principles of Geology, Being an Attempt to Explain the Former Changes of the Earth's Surface, by Reference to Causes Now in Operation*, 3 vols. University of Chicago Press, Chicago.

A facsimile edition of this 1830-1833 treatise, which introduced the public to geological gradualism and greatly influenced Darwin who had volumes 1 and 2 with him on his epic voyage of the *Beagle*.

Malthus, Robert. 1970. *An Essay on the Principle of Population, as it Affects the Future Improvement of Society*. Penguin Classics, New York.

Reprint of original 1798 edition. Reacting to the Enlightenment's utopian perfectionism, Malthus painted a darker picture of famine and disease. His conclusion that population growth is "indefinitely greater than the power in the earth to produce subsistence for man" was a key concept for Darwin in developing his theory of Natural Selection.

Paley, William. *Natural Theology; or Evidence of the Existence and Attributes of the Deity*. 2006. Oxford University Press, New York.

A reprint of Paley's 1802 classic defense of intelligent design.

ACKNOWLEDGMENTS

Production of this book would have been impossible without the help of many members of the staff of the American Museum of Natural History. For allowing us access to collections under their care, for aid in identification, and for many other kindnesses, we thank: Department of Anthropology: Paul Beelitz, Jaymie Brauer, Joe Jimenez, Gary Sawyer, Ian Tattersall and David Hurst Thomas. Department of Invertebrates: Melvin Hinkley, Sidney S. Horenstein and Neil Landman. Department of Vertebrate Paleontology: John Alexander, Bob Evander, Charlotte Holton, Jeanne Kelly, Bryn Mader, John Maisey, Malcolm McKenna, Mark Norell and Ivy Rutzky.

Thanks, too, to Nicholas LiVolsi, for an elegant book design; Carl Zimmer for his eloquent introduction; George Dick and Jared Stevens, Four Colour Print Group for valuable print production advice; and Lionel Koffler, Michael Worek, Jacqueline Hope Raynor, Diane Vanderkooy and Tom Martin, Firefly Books for expert editorial and publishing support.

Special thanks to Peter N. Névraumont: he conceived this book, prodded its author and photographer into timely completion, and spent hours uncountable seeing it through the editorial and production process. And to the late Ann Julia Perrini, who as President of Névraumont Publishing Company kept the whole original operation afloat, making this updated and revised edition possible.

PLATE 160

PALAEOPYTHON FISCHERI,
A TREE DWELLING BOA.

Middle Eocene (*ca*. 47 million years).
Messel Shales, Eichstatt, Bavaria.

Darwin had concluded that evolutionary changes generally happen incrementally and very slowly over millions of years. In contrast, my research has convinced me that the pace of evolution is typically episodic and that the appearance of new life forms follows extinction events.

Credit: Senckenberg Forschungsinstitut und Naturmuseum.

INDEX